AF581705

The Contribution of Food Oral Processing

The Contribution of Food Oral Processing

Editors

Susana Fiszman
Amparo Tarrega

MDPI • Basel • Beijing • Wuhan • Barcelona • Belgrade • Manchester • Tokyo • Cluj • Tianjin

Editors
Susana Fiszman
Instituto de Agroquímica y Tecnología de Alimentos (IATA-CSIC)
Spain

Amparo Tarrega
Institute of Agrrochemistry and Food Technology IATA- CSIC
Spain

Editorial Office
MDPI
St. Alban-Anlage 66
4052 Basel, Switzerland

This is a reprint of articles from the Special Issue published online in the open access journal *Foods* (ISSN 2304-8158) (available at: https://www.mdpi.com/journal/foods/special_issues/food-oral).

For citation purposes, cite each article independently as indicated on the article page online and as indicated below:

LastName, A.A.; LastName, B.B.; LastName, C.C. Article Title. *Journal Name* **Year**, *Article Number*, Page Range.

ISBN 978-3-03936-834-1 (Hbk)
ISBN 978-3-03936-835-8 (PDF)

Contents

About the Editors

Susana Fiszman holds a PhD in Chemistry from the University of Valencia; she is a Research Professor (Professorial Fellow) of the Spanish National Research Council (CSIC). She developed her professional career at the Institute of Agrochemistry and Food Technology (IATA) in Valencia, Spain. She is currently a member of several prestigious international expert panels, and is a member of the Editorial Board of a number of journals in the Food Science and Technology field. She has authored over 200 scientific publications (research papers, reviews, and book chapters). She also actively collaborates with researchers of many international groups. She has dedicated her professional life to the study of food texture properties and their sensory perception, focusing on how to create and modify food structures. She collaborates regularly on editing books in her subject area and participates actively in national and international conferences in her field of expertise. A specific area in which she has worked in recent years is the relationship between food oral processing and expected perceptions of satiety and satiation, since these sensations are directly related to how food is processed in the mouth. Her studies will assist in defining strategies to design products that meet the needs of specific consumer groups directly related to their health and wellbeing.

Amparo Tarrega is a Researcher in Sensory and Consumer Science at the Instituto de Agroquímica y Tecnología de Alimentos (CSIC, Spain). She has developed part of her research in renowned institutions such as Cornell University (USA), INRA Dijon (France), and the University of Nottingham (United Kingdom). Her research work focuses on three main lines: 1) Advancing knowledge in texture and taste perception mechanisms; 2) Instrumental methods to predict sensory properties; and 3) The application of new methods in sensory food assessment and consumer response.

Editorial

Oral Processing Studies: Why Multidisiciplinary?

Susana Fiszman * and Amparo Tarrega

Instituto de Agroquímica y Tecnología de Alimentos (IATA-CSIC), 46980 Valencia, Spain; atarrega@iata.csic.es
* Correspondence: sfiszman@iata.csic.es

Received: 15 June 2020; Accepted: 2 July 2020; Published: 3 July 2020

Abstract: When food is ingested, it remains in the mouth for a short period of time. Although this period is brief compared to the total food nutrient digestion and absorption time, it is crucially important as it is the first step in digestion. It is also very important that, while the food is in the mouth, it is perceived by the senses and then a decision is made on swallowing. Oral sensory perception is an integrative response, which is generated in very short time (normally a few seconds) from complex information gathered from multiple sources during mastication and swallowing. Consequently, food oral processing studies include many orientations. This Special Issue brings together a small range of studies with a diversity of approaches that provide good examples of the complexity and multidisciplinarity of the subject.

Keywords: oral processing; mastication; swallowing; temporal dynamic measurements; in-mouth sensory perception

---

1. Introduction

From the moment a piece of food enters the mouth, it is continuously assessed to decide how to handle it in the next instant. This depends on the level of force at which it breaks, how it breaks, how the broken pieces behave, how many and how big they are, how much saliva they absorb, how easily they form a bolus, among other factors related to the series of actions which lead to the final swallow. Throughout this time span (only a few seconds), the senses of touch, smell, taste, and hearing collect data to form an integrative perception of that food.

Consequently, oral processing can be approached from several points of view, such as physiological, nutritional, psychological, metabolic, sensory, mechanical, and chemical, among others and, importantly, all of them can be studied with a sense of evolution and a sense of dynamics.

One study in the present themed issue adopted an original sensory approach to wine [1]. It focused on "body", a quality of wine that could be roughly defined as "how it feels inside the mouth", which is constantly mentioned by experts, sommeliers, and consumers. A number of individual components of wine have previously been associated with "body" and their instrumental analysis has been correlated to sensory perception. In this study, several of these components and their combinations were tested [1], introducing saliva as a new variable, and instrumental and sensory analyses were performed and correlated to wine body perception.

An interesting mechanical approach was taken in an investigation of food for a population of advanced age [2]. Care food is normally designed to be broken easily in the mouth just by pressing it against the palate, without the participation of teeth. To follow the fracture patterns of the experimental food, the authors used an artificial tongue made of a soft material instead of the hard surfaces of a conventional texture analyzer. Video recordings of the food deformation and fracture indicated that the soft system mimics the natural oral processing of care food in a more reliable way.

Patterns of in vitro fragmentation were also studied in another paper [3] in the present themed issue, but through a different approach. In this case, the influence of the food matrix (rather than the composition) was investigated with oat biscuits. In vitro fragmentation of biscuits with the same composition differed depending on the morphology of the oat ingredient. In turn, these differences play a role in the level of starch hydrolysis in the oral phase.

Temporal aspects of oral perception dominate the three remaining papers. One studies the relative importance of several factors in the temporal aroma-compound release that takes place during eating [4]. These factors are the physicochemical properties of the food matrix, the characteristics of the aroma compounds, and oral physiological parameters, such as bite force, shearing angle, and salivary flow rate. All the oral parameters were measured using a chewing simulator and the aroma compound releases were followed in real time through atmospheric pressure chemical ionization mass spectrometry. Interesting differences in aroma release were found depending on the hydrophobicity of the volatile compound.

The second paper, dealing with the temporal aspects of food oral processing, studies the in-mouth perception of carrot purees designed for dysphagic patients, which were thickened with starch and xanthan gum, and their combinations [5]. Two temporal sensory techniques were used: Temporal Dominance of Sensations (TDS); Temporal Check-all-that-Apply questions (TCATA). "Grainy" was associated with the starch-thickened puree, while "smooth" and "slippery" were associated with xanthan gum. The oral perception of all the puree samples evolved from predominantly thick to adhesive, with a mouthcoating sensation towards the end of the evaluation. Both TDS and TCATA yielded similar results.

Finally, the third of these papers studies the temporal perception of the texture of complementary porridges for infants and young children used in low-income African communities [6]. This paper is very interesting not only technically but also because of its social dimension. The sensory quality of these porridges affects oral abilities, leading to malnutrition. The TCATA sensory method was used to assess indigenous and commercial porridges. The assessors used two different oral techniques: up and down movements to mimic babies' mastication; normal movements, as in normal eating. The results reveal that the indigenous porridges were too thick and sticky, and not easy to swallow, even with a low solids content—especially using the Up-Down method. This highlights that texture improvement should be promoted for this complementary type of feeding.

The diversity of the approaches presented in "Contribution of Food Oral Processing" encourages an exploration of potential areas of collaboration which could give rise to a better understanding of the mechanisms involved in eating. The editors hope the readers will find this issue interesting and useful for inspiring future research.

Author Contributions: Authors S.F. and A.T. have equally contributed to the visualization, writing, review and editing the present Editorial. All authors have read and agreed to the published version of the manuscript.

Conflicts of Interest: The authors declare no conflict of interest.

References

1. Laguna, L.; Álvarez, M.D.; Simone, E.; Moreno-Arribas, M.V.; Bartolomé, B. Oral Wine Texture Perception and Its Correlation with Instrumental Texture Features of Wine-Saliva Mixtures. *Foods* **2019**, *8*, 190. [CrossRef] [PubMed]
2. Kohyama, K.; Ishihara, S.; Nakauma, M.; Funami, T. Compression Test of Soft Food Gels Using a Soft Machine with an Artificial Tongue. *Foods* **2019**, *8*, 182. [CrossRef] [PubMed]
3. Tarrega, A.; Yven, C.; Semon, E.; Mielle, P.; Salles, C. Effect of Oral Physiology Parameters on in-Mouth Aroma Compound Release Using Lipoprotein Matrices: An in Vitro Approach. *Foods* **2019**, *8*, 106. [CrossRef] [PubMed]
4. Gamero, A.; Nguyen, Q.-C.; Varela, P.; Fiszman, S.; Tarrega, A.; Rizo, A. Potential Impact of Oat Ingredient Type on Oral Fragmentation of Biscuits and Oro-Digestibility of Starch—An in Vitro Approach. *Foods* **2019**, *8*, 148. [CrossRef] [PubMed]

5. Sharma, M.; Duizer, L. Characterizing the Dynamic Textural Properties of Hydrocolloids in Pureed Foods—A Comparison between TDS and TCATA. *Foods* **2019**, *8*, 184. [CrossRef] [PubMed]
6. Makame, J.; Cronje, T.; Emmambux, N.M.; De Kock, H. Dynamic Oral Texture Properties of Selected Indigenous Complementary Porridges Used in African Communities. *Foods* **2019**, *8*, 221. [CrossRef] [PubMed]

Article

Oral Wine Texture Perception and Its Correlation with Instrumental Texture Features of Wine-Saliva Mixtures

Laura Laguna [1,2,*], María Dolores Álvarez [3], Elena Simone [4], Maria Victoria Moreno-Arribas [1] and Begoña Bartolomé [1]

1 Institute of Food Science Research (CIAL), CSIC-UAM, 28049 Madrid, Spain; victoria.moreno@csic.es (M.V.M.-A.); bartolome@ifi.csic.es (B.B.)
2 Institute of Agrochemistry and Food Technology (IATA), 46980 Paterna, Spain
3 Leeds, Institute of Food Science, Technology and Nutrition (ICTAN- CSIC), 28040 Madrid, Spain; laura.laguna@csic.es
4 Food Colloids and Processing Group, School of Food Science and Nutrition, University of Leeds, Leeds LS2 9JT, UK; E.Simone@leeds.ac.uk
* Correspondence: laura.laguna@csic.es; Tel.: +34-963-900-022

Received: 23 April 2019; Accepted: 30 May 2019; Published: 1 June 2019

Abstract: Unlike solid food, texture descriptors in liquid food are scarce, and they are frequently reduced to the term viscosity. However, in wines, apart from viscosity, terms, such as astringency, body, unctuosity and density, help describe their texture, relating the complexity and balance among their chemical components. Yet there is uncertainty about which wine components (and their combinations) cause each texture sensation and if their instrumental assessment is possible. Therefore, the aim of the present work was to study the effect of wine texture on its main components, when interacting with saliva. This was completed by using instrumental measurements of density and viscosity, and by using two types of panels (trained and expert). For that, six different model-wine formulations were prepared by adding one or multiple wine components: ethanol, mannoproteins, glycerol, and tannins to a de-alcoholised wine. All formulations were mixed with fresh human saliva (1:1), and their density and rheological properties were measured. Although there were no statistical differences, body perception was higher for samples with glycerol and/or mannoproteins, this was also correlated with density instrumental measurements ($R = 0.971$, $p = 0.029$). The viscosity of samples with tannins was the highest due to the formation of complexes between the model-wine and salivary proteins. This also provided astringency, therefore correlating viscosity and astringency feelings ($R = 0.855$, $p = 0.030$). No correlation was found between viscosity and body perception because of the overlapping of the phenolic components. Overall, the present results reveal saliva as a key factor when studying the wine texture through instrumental measurements (density and viscosity).

Keywords: wine texture; body; viscosity; density; trained and expert panel

1. Introduction

Food texture is a sensory property; it is defined as "the sensory and functional manifestation of the structural, mechanical and surface properties of foods detected through the senses of vision, hearing, touch and kinaesthetics" [1]. In liquid food, the perceived texture vocabulary is rather scarce, being often reduced to the term viscosity [2]. In wines, the perceived texture is referred to as 'mouthfeel' [3] and arises from the changes induced by the wine to the integrity of the salivary film that coats oral mucosa. These changes are perceived by the free nerve endings (also called tactile sensors) located in the connective tissue of filiform papillae in the mouth [4]. In wines, because of the complexity conferred by its components, describing the perception of texture often involves several terms, such as astringency, body [5], unctuosity, density and viscosity [6].

Sensory and instrumental approaches could contribute knowledge about wine texture or mouthfeel. From a sensory point of view, wine tasting has an enormous tradition being generally assessed by 'wine experts'. Wine experts include oenologists and sommeliers, among others; and their expertise allows them to award wine quality. These specialists often influence the average wine consumer as the consumers follow the experts' awarded medals or brands, quality lists and opinion articles [7]. However, from a scientific perspective, there is no control in expert panels performance as there is uncertainty about what sensory attributes are being analysed and what their definitions are. In sensory science, for the need to control the variables as much as possible, trained panels are normally used to perform sensory descriptive analysis (QDA) [8]. QDA results are generally used to correlate sensory and instrumental measurements. Yet, for wine, finding these correlations and relating instrumental or sensory sensations to the presence or content of the principal chemical components remains a challenge.

The wine components most cited for influencing wine texture sensations include ethanol [9], phenolic compounds [10–12], glycerol [13,14] and polysaccharides [15,16]. The alcohol content in wines is often between 11–14 mL/100 mL, although it can be as low as 7.5 mL/100 mL in some botrytised wines, or 15 mL/100 mL and above in some red and dessert wines. In early research, there was a suggestion that ethanol played an important role in wine body [17]. However, in later studies, it was discovered that ethanol had little or no effect on body perception or viscous mouthfeel [18–21]. Among wine polyols, the most significant is glycerol, with concentrations reaching 10 g/L in red wines and 7 g/L in white wines. From a sensory point of view, attributes associated with glycerol are oiliness, persistence and mellowness. Still, glycerol has no detectable effect in in-mouth viscosity below a concentration of 25 g/L [14]. The total phenolic content (among other composition factors) of wine depends on many factors (e.g., grape variety, growing conditions, harvest time and/or the winemaking process) [22]. Total content can vary between 40–400 mg/L in white wines, 900–1400 mg/L in young red wines and 1600–2500 mg/L in aged red wines [23]. From a sensory perspective, controversy exists regarding the size of polyphenols or phenols structure and the astringency intensity perception. Classically, it is accepted that differences between phenolic compounds produce different saliva protein precipitation. Polyphenols with an extended structure have a higher affinity to proline-rich proteins (PRPs) as the number of interaction sites increases with polyphenols size, promoting protein precipitation [24]. But previous work [10] showed that, independently of the chemical affinity and structures, when using a trained panel, the same levels of astringency was perceived, among different aqueous-polyphenol mixtures. The only difference found was that catechins were perceived as slightly bitter. Still, a later study [25] showed that the average particle size of flavanols and saliva complexes (measured by light scattering) increased with its concentration and was linked with an increment in astringency perception.

This controversy might be caused by various factors. One of them is the bitterness, felt alongside astringency and associated with polyphenols. Therefore, the difference perceived in astringency might be linked with taste perception and not with mouthfeel [26]. Another factor is the phenols lingering presence in the mouth. A recent study [27] showed that the phenolic component stayed in the mouth for more than 2 minutes. This can cause cross-mouth-contamination between samples compromising the validity of the sensory results. Finally, in wine perception, there are other macromolecular fractions, such as polysaccharides, that have been reported to influence the texture perception of wine. These components derive from cell walls of yeast (mannoproteins from *Saccharomyces cerevisiae*) [28], grapes cells (arabinogalactan-proteins) [29] and other sources (e.g., bacteria). A mixture of neutral polysaccharides (mannoproteins and arabinogalactan-protein complexes) and acidic polysaccharide (rhamnogalacturonan II) significantly increases the 'fullness' sensation, above that of the base wine using QDA [30].

In summary, previous scientific works have studied the relationship of 'body' with ethanol glycerol or polysaccharides. Still, there is no work that integrates these components in a model using human physiological conditions, such as integrating saliva, that, until now, has only been studied in terms of astringency perception. The study of wine components joined with saliva can help wineries to find instrumental measurements to adjust wine components for specific mouthfeel characteristics.

All things considered, the objective of the present study was twofold. First, to investigate instrumentally, by measuring viscosity and density, the effect of principal wine components classically related to wine mouthfeel perception (ethanol, mannoproteins, glycerol and tannins) in model based-wines (de-alcoholised) mixed with saliva. Second, to investigate the correlations of these instrumental results with human perception by using two types of panels (trained and expert). For that, artificial model-wines with one or multiple ingredients were created. It was mixed with human saliva at 37 °C, and the resulting mixture was measured for its instrumental properties (apparent viscosity and density). Then, two panels performed sensory analysis, one trained in wine mouthfeel attributes and one expert wine panel.

2. Material and Methods

2.1. Materials

2.1.1. Model-Wine Samples

A commercial white de-alcoholised wine (0.5 mL/100 mL alcohol content) was used as a base wine in all formulations (Torres Natureo Muscat 2014, Miguel Torres S.A. Winery, Barcelona, Spain). Preparation of model-wine samples (Table 1) were formulated with either the presence or absence of ethanol (E) (ethanol absolute food grade, AppliChem, Panreac, Barcelona, Spain), yeast mannoproteins (M) (Mannoplus, Agrovin, S.A. Ciudad Real, Spain), glycerol (G) (Mineral Waters, Purfleet, UK) and tannins (T) (oak tannin, Agrovin, S.A., Ciudad Real, Spain). The concentration of the different components was chosen based on their quantities in commercial red wines, except for ethanol, which was chosen to comply with a minimum legal alcohol content of wine (8 mL/100 mL). In initial experiments, a higher content of alcohol was chosen (14%), but as this was added pure, it resulted in overpowering the senses.

Table 1. Model-wine samples made of the base wine (W), ethanol (E), mannoprotein (M), glycerol (G) and tannins (T).

Sample	Base Wine (mL)	Ethanol (mL)	Mannoproteins (g)	Glycerol (g)	Tannins (g)
W	100	-	-	-	-
WE	92	8	-	-	-
WM	100	-	0.5	-	-
WG	100	-	-	1	-
WT	100	-	-	-	0.1
WEMGT	92	8	0.5	1	0.1

All ingredients used were food grade and dissolved/dispersed in the commercial de-alcoholised wine base. Samples were prepared the day before analysis and sensory evaluation.

2.1.2. Saliva/Sample Mixtures

Fresh human saliva (10 mL) was provided by a healthy volunteer (23 years old). Immediately after collection, the saliva was mixed with each model-wine formulation in a proportion 1:1 (*v/v*) only for instrumental measurements.

2.2. Methods

2.2.1. Physical Measurements in Saliva/Sample Mixtures

Density

A digital densitometer was used (Anton Paar density metre, Graz, Austria) at 37 °C, measuring the density of the different wine formulations and their mixtures with human saliva (see Section 2.2). Tests were performed in triplicate for each sample.

Flow Behaviour

Rheological tests were conducted on the mixtures of model-wine formulations, model-wine formulations with human saliva and water with saliva. The shear rate ($\dot{\gamma}$) was measured in a rotational Kinexus pro rheometer (Malvern Instruments Ltd., UK), equipped with a 40 mm cone (1°) and a plate geometry with a gap of 0.15 mm. Five hundred microlitres of each mixture were placed with a pipette onto a pre-heated plate (37 °C). Control of the temperature, to ± 0.1 °C, was by a Peltier element in the lower plate. A cover was used to maintain the temperature (37 °C) and to avoid evaporation of the samples.

Flow curves were obtained at a shear rate ($\dot{\gamma}$) ranging from 0.1 to 100 s^{-1}, selected based on previous studies [31,32]. The resulting flow curves represent viscosity as a function of shear rate, and obtained results were fitted to the Ostwald de Waele model ($\eta = K\dot{\gamma}^{n-1}$). K (Pa s) is the consistency index, and n is the flow index. The least-square data fit was good (R^2 > 0.900) in all cases. Comparisons of samples were made using apparent viscosity values at a shear rate of 100 s^{-1}. Each sample was measured in triplicate.

2.2.2. Wine-Oral Texture Evaluation

Evaluation by Trained Panel

A panel of 8 assessors (5 women and 3 men, 20–34 years old) with one year of training in wine mouthfeel characteristics took part in this study. In a previous trial, their sensory thresholds in viscosity, astringency and alcoholic content in wine were tested [33].

In a preliminary session, the panellists were asked to generate the attributes perceived, relating to the mouthfeel characteristics of the six samples (Table 2). If they found some specific taste or aroma attribute key for wine discrimination, they were encouraged to record the feeling. Then, they agreed on the attributes to be chosen and their definitions (Table 2).

Table 2. Sensory attributes selected by the trained panel and its consensus definitions.

Term	Definition
Body	Viscosity sensation when swishing
Astringency	Dryness from the tongue tip to the throat
Alcoholic feeling	Hot sensation typical in alcoholic beverages
Cereal taste	Feeling of cereal taste
Bitter taste	Bitter taste at the end of the tongue

In the first two sessions (around 30 minutes), the characteristics were elicited and discussed for samples W and WEMGT, as they were the most opposed of the sample set. In the following sessions, review for the commonality of all samples provided and attributes previously generated helped compilation of a final list of attributes, as shown in Table 2. The creation of their own language was under the supervision of the panel leader, who ensured that attributes were non-redundant towards samples. The aim of these sessions was for all panellists to use the same concepts allowing them to communicate precisely with each other. Finally, along with tactile (mouthfeel) attributes, two taste attributes were also included in the final attribute list (cereal taste and bitter taste).

For the formal assessment, three sessions on different days took place. Samples (20 mL) were presented monadically in separate wine glasses, each labelled with 3-digit random codes. All tests conducted were at wine serving temperature (14–17 °C). Panellists performed the test in a room isolated from odours and noise. They were advised to wait at least one minute between samples, while water, crackers and carrots were offered as palate cleansers; the six samples were tested in triplicate. The panellists were then asked to score the attributes (Table 2) at two times, during the consumption and after 10 seconds from swallowing the sample. They used a 10 cm unstructured scale (anchored from weak to strong) to score the intensity of the attributes (body, astringency, alcoholic feeling, cereal taste and bitter taste) of the samples.

Evaluation by Expert Panel

Nine wine experts were recruited to take part in a blind tasting of the wines. Experts included oenology teachers, oenology students and other wine professionals (winery oenologists). The formal evaluation was made in the same way as the trained panel, including the same number of sessions, samples per session and sample order. No training sessions were performed with this panel.

2.3. *Data Analysis*

Analysis of variance (one-way ANOVA) was applied to study the differences between the wine samples in the sensory analysis scores, density and rheology values. Tukey test was used for post hoc mean comparisons at a 95% significance level ($p < 0.05$). Paired-samples T-tests were conducted to compare the attribute scores of the two panels.

Pearson's correlation between the instrumental analyses results and the mean intensity scores from the sensory descriptive test by the trained panel were applied.

All the statistical tests were done using IBM SPSS Statistics for Windows, Version 24.0. (IBM Corp., Armonk, NY, USA).

3. Results

3.1. *Instrumental Measurements*

3.1.1. Density

Figure 1 shows the density measurements of model-wine mixed with saliva. There were no statistical differences among samples of model-wine with saliva. However, it can be observed that the two samples that contained ethanol, WE and WEMGT, had lower density values. This was expected since the density of ethanol is lower than water (water σ at 20 °C = 1 g cm^{-3}, ethanol σ at 20 °C = 0.791 g cm^{-3}). Although not significant, the sample with glycerol (WG) had a higher density, followed by the sample with mannoproteins (WM) and tannins (WT). These samples with higher density also had less variability among measurements. Authors believed that the previous ingredients (glycerol, mannoproteins and tannins) helped the saliva-wine bonding stabilisation. Previous work, with the aim of understanding body perception, measured wine density [34]. It was found that density was related to the alcohol content, and in full-bodied wines, with the alcohol content and dry extract. Besides alcohol content, the perception of the body in wines has also been linked with mannoproteins [30] content and glycerol [19]. Although in previous studies, no saliva was added to wine, the results obtained in this work followed the same trend, in the way that in the presence of glycerol and mannoproteins, density increased.

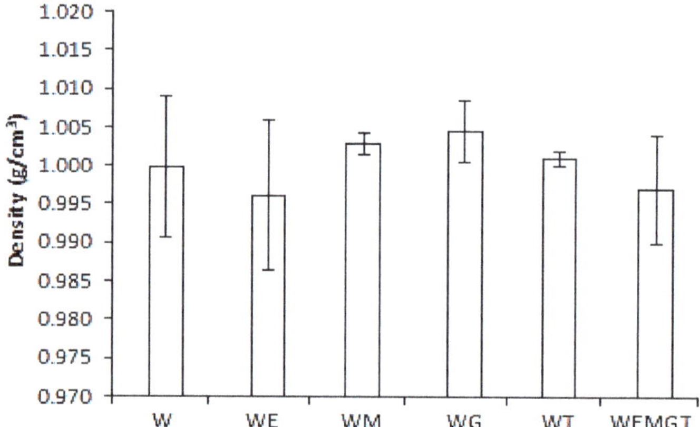

Figure 1. Density measurements of model-wine formulations with saliva. Bars represent the standard deviation. Model-wine letters indicate components: base wine (W), ethanol (E), mannoprotein (M), glycerol (G) and tannins (T).

3.1.2. Flow Behaviour

Although there is no agreement for the shear rate with which wine undergoes inside the mouth, the classical master curve by Shama and Sherman [31] showed that deformation of food in the mouth would be at shear rates that oscillate between 10 and 1000 s^{-1}. On the other hand, as food viscosity increases, the tongue exerts higher forces; therefore, for liquids, shear rates will be lower ($\approx 100\ s^{-1}$) [32]. In the present study, the flow curve obtained was between 0.1–100 s^{-1}. A shear of 100 s^{-1} was considered as the value for tongue against the palate in physiological conditions when drinking. The apparent viscosity at this value was calculated to compare the different model-wine samples with saliva (Table 3). The flow curves are presented in Figure 2.

Table 3. Apparent viscosity (Pa s) of model-wines at a shear rate of 100 s^{-1}.

Model-Wine Samples	η (at $\dot{\gamma}$ 100 s^{-1})
W	0.0167 ± 0.002 ^{ab}
WE	0.0110 ± 0.001 ^{a}
WM	0.0200 ± 0.002 ^{ab}
WG	0.0117 ± 0.002 ^{ab}
WT	0.0347 ± 0.002 ^{c}
WEMGT	0.0203 ± 0.002 ^{b}

Same letters indicate no significant differences among samples ($p < 0.05$). Model-wine letters indicate components: base wine (W), ethanol (E), mannoprotein (M), glycerol (G) and tannins (T).

All samples, including a control sample (water plus saliva) for comparative purposes, showed a shear thinning behaviour as apparent viscosity dropped when the shear rate increased. The sample WT was significantly the most viscous, followed by the other tannin-containing sample (WEMGT). This result agreed with a previous study [13] where instrumental viscosity increment resulted from the formation of a saliva protein-polyphenol complex. This complex has a higher hydrodynamic diameter, shown by using dynamic light scattering and negative-stain transmission electron microscopy. Formation of these complexes is via hydrogen bonding between hydroxyl groups of phenolic compounds and carbonyl/amide groups of proteins. Hydrophobic interactions between the benzoic ring of phenolic compounds and the apolar side chains of amino acids, such as leucine, lysine and proline, in salivary proteins [11,35] also contributed to this complex formation.

Samples with mannoproteins (WM) showed lower viscosity values than samples with tannins (WT). Mannoproteins are macromolecular fractions, present in wines generally used to stabilise wine flavour, colour and foam (in sparkling wines) [36]. In this case, it is believed that the viscosity increase is due to the size of the molecules in the wine mixture. When mannoproteins and tannins were present in the same wine (WEMGT), the viscosity was slightly lower because of their influence on the tannins size [37], as mannoproteins act by encapsulating polyphenols, thus interfering in the protein binding [38].

The least viscous samples were WE and WG, with no significant differences among them. Previous authors, using an increasing quantity of ethanol and glycerol, investigated their relationship with the contribution of 'body' in wine [39]. Authors found that in ranges of ethanol content 0–15% (*v/v*) and glycerol 0–20 g/L, the viscosity varied linearly; for every 1% increase in ethanol concentration, viscosity increased by 0.047×10^{-3} mPa s; for every g/L increase in glycerol concentration, viscosity increased by 0.005×10^{-3} mPa s. However, authors only measured the wine samples without the addition of saliva, and not considering its interaction. Further to this, in this present work, the ranges of ethanol and glycerol added were smaller. Therefore, the addition of saliva and its diluting effect, plus the quantity of glycerol and ethanol added, had led to non-significant changes to WE and WG samples.

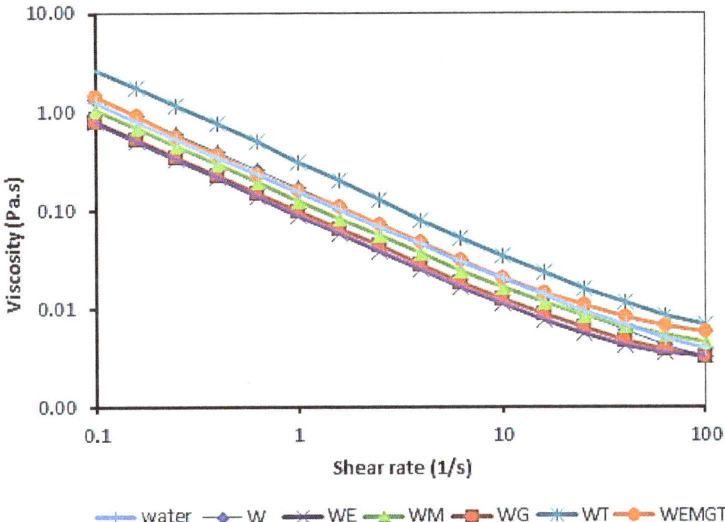

Figure 2. Dynamic viscosity of model-wine formulations with human saliva (1:1). W (♦), WG (■), WM (▲), WEMGT (●), WE (X), WT (X), water (l). Model-wine letters indicate components: base wine (W), ethanol (E), mannoprotein (M), glycerol (G) and tannins (T).

3.2. Descriptive Sensory Evaluation

3.2.1. Wine-Oral Sensations

Table 4 shows the mean scores of sensory attributes in the mouth (4a) and after swallowing (4b) for both trained and expert panels. Descriptive sensory techniques can be used when searching for sensory-instrumental relationships [8]; therefore, first, results from the trained panel have been analysed.

Table 4. Mean of descriptive sensory scores in the mouth for trained panel and expert panel for wine-in-mouth sensations when (**a**) consuming the samples and (**b**) 10 seconds after swallowing.

	Trained Panel					Expert Panel				
	Body	Astringency	Alcohol	Cereal	Bitter	Body	Astringency	Alcohol	Cereal	Bitter
(a)										
W	2.71 ^{b*}	1.40 ^{b}	2.84 ^{bc}	0.52 ^{a}	0.56 ^{b}	3.62 ^{b}	2.04 ^{b}	1.07	2.46 ^{c}	1.71 ^{c}
WE	3.53 ^{ab}	2.61 ^{ab}	4.99 ^{a}	0.96 ^{a}	4.07 ^{a}	4.96 ^{a}	3.73 ^{ab}	4.56	2.59 ^{c}	4.30 ^{a}
WM	4.09 ^{a}	2.92 ^{ab}	2.36 ^{c}	1.25 ^{a}	1.42 ^{b}	4.18 ^{ab}	2.73 ^{ab}	1.46	3.06 ^{bc}	1.91 ^{c}
WG	4.57 ^{a}	3.50 ^{ab}	2.08 ^{c}	0.81 ^{a}	1.18 ^{b}	4.33 ^{ab}	2.88 ^{ab}	2.31	2.298 ^{c}	1.69 ^{c}
WT	3.60 ^{ab}	5.28 ^{a}	2.37 ^{c}	1.65 ^{a}	2.15 ^{ab}	3.92 ^{ab}	4.28 ^{a}	2.56	4.21 ^{ab}	2.84 ^{b}
WEMGT	4.02 ^{a}	4.369 ^{ab}	3.54 ^{b}	1.71 ^{a}	3.89 ^{a}	4.77 ^{a}	3.88 ^{ab}	4.03	4.52 ^{a}	3.43 ^{b}
(b)										
W	1.34 ^{a}	1.33 ^{a}	0.34 ^{a}	0.67 ^{a}	0.68 ^{a}	3.50 ^{a}	3.00 ^{a}	1.67 ^{a}	2.00 ^{a}	1.12 ^{a}
WE	2.01 ^{a}	3.00 ^{a}	3.43 ^{a}	1.00 ^{a}	2.33 ^{a}	4.00 ^{a}	3.70 ^{a}	4.00 ^{a}	2.33 ^{a}	3.33 ^{a}
WM	2.00 ^{a}	3.08 ^{a}	0.66 ^{a}	1.57 ^{a}	1.00 ^{a}	3.60 ^{a}	2.67 ^{a}	2.00 ^{a}	2.67 ^{a}	1.67 ^{a}
WG	2.05 ^{a}	1.98 ^{a}	1.00 ^{a}	1.65 ^{a}	1.15 ^{a}	3.23 ^{a}	2.33 ^{a}	1.70 ^{a}	1.80 ^{a}	1.67 ^{a}
WT	2.37 ^{a}	4.02 ^{a}	0.69 ^{a}	1.00 ^{a}	2.00 ^{a}	3.31 ^{a}	4.00 ^{a}	2.20 ^{a}	3.50 ^{a}	1.95 ^{a}
WEMGT	2.67 ^{a}	3.68 ^{a}	2.43 ^{a}	2.00 ^{a}	3.00 ^{a}	3.98 ^{a}	3.68 ^{a}	3.00 ^{a}	4.00 ^{a}	2.67 ^{a}

Model-wine letters indicate components: base wine (W), ethanol (E), mannoprotein (M), glycerol (G) and tannins (T).
* Tukey test among wine-model formulations, same letter in the same column, does not differ significantly, $p < 0.05$.

Significant differences among the model-wines were found between their body and intensity perception ($p < 0.05$). Average body perception ranged from 2.71 (W: control) to 4.57 (WG: wine with glycerol), as shown in Table 4. Model-wine containing mannoproteins (WM) and model-wine containing mannoproteins and glycerol (WEMGT) were also scored with high intensity, with no significant difference from WG.

Both components (mannoproteins and glycerol) have been previously reported as influencers in wine mouthfeel [18,19]. Mannoproteins are composed of polypeptides and linked to highly branched mannose side chain by glycosidic bonds. This big molecule (800,000 Daltons) has been previously linked with a 'fullness' sensation when using a trained panel [30]. Present authors believe that because of the size of this molecule, the wine acquires a structure that is perceived as bulkier mouth feeling. Therefore, the trained panel considered this wine sample as with more body.

In the case of glycerol, the previous bibliography has not linked it specifically to 'body'. It has been linked with other various attributes, such as oiliness, persistence and mellowness [14]. In a tribological study, it was found to decrease the friction coefficient [13]. However, the definition of the body itself is still a pending subject. In fact, there is no consensus in the reference material to be used and can include dairy products, such as whipping cream [40], gels [19] or xanthan gum [41].

Astringency perceived was higher for samples with added oak tannin (T). The added tannins formed complexes with the salivary protein; these complex formations have been previously described by two different bond types: hydrogen bonding (between hydroxyl groups of phenolic compounds and carbonyl/amide group of protein) and by hydrophobic bonding (between the benzoic ring of phenolic compounds and the apolar side chains of the salivary protein's amino acids) [35]. This leads to the formation of a complex caused by proteins in saliva precipitating. Therefore, the salivary film that covers the oral mucosa loses its structure and separates from the oral mucosa causing the mouth to become dry [24]. When mannoproteins and ethanol are present, a saliva-polyphenol complex formation explains their interference and a decrease in the perception of astringency [35,36]. Other samples with the same quantity of tannins showed less astringency. In a previous study, authors found that hydrodynamic diameter of saliva with tannins increased (by a factor of almost 2.5–3); but in the presence of ethanol or glycerol, it decreased [13]. This might be due to the ability of ethanol to interfere with wine polyphenol-PRPs interactions [42]. Further, mannoproteins interferes in the tannin aggregation [37] by three mechanisms: i) encapsulating polyphenols and interfering with their ability to bind proteins [38], ii) enhancing polyphenols solubility in an aqueous medium through the form of protein-polyphenol aggregates [38], and iii) by binding salivary proteins, avoiding the polyphenols attachment [26].

In addition, as it can be seen in Table 4, for the trained panel, other samples not containing tannins also contributed to the astringency feeling. This is because, in wine, astringency is also produced by the low pH of the samples [43–45] due to salivary protein precipitation.

Ethanol perception was inevitably higher for the sample containing ethanol (WE). Ethanol is an effective sensory stimulant, activating brain gustatory circuits, as well as trigeminal pathways, sensitive to an irritant stimulus [46], providing sensations of hotness and bitterness that linger in mouth [47,48]. Besides alcoholic feeling, ethanol presence also produced a significant bitterness perception. In a more complex wine-model matrix, the alcohol feeling was shown to be decreased (WEMGT), and this was caused by its interaction with tannins [49]. Another reason could be the sensory perception displacement effect when presenting a different stimulus at the same time.

Besides its initial sensory profile with the trained panel, the attribute means were analysed for significant differences ($p < 0.05$) between panels (paired comparison) for each sample, and between samples within each panel (Tukey's test). There were significant differences ($p < 0.05$) in the perception of 'body' between panels for the samples containing ethanol. The expert panel found ethanol to be the most influential component in body perception (WE = 4.96 ± 1.63, WEMGT = 4.77 ± 1.36) whilst the trained panel perceived 'body' significantly higher for samples containing glycerol, mannoproteins or both (WG = 4.56 ± 1.5, WM = 4.09 ± 1.5, WEMGT = 4.01 ± 2.1). Previous work [28] agrees with the results of both panels, as all of three components, glycerol, mannoproteins and ethanol, are influencers on the wine body [28].

For 'astringency', there were significant differences ($p < 0.05$) between panels, as the trained panel scored the astringency significantly higher (Table 4a). Both panels found the sample WT significantly more astringent, followed by WEMGT. Therefore, here, tannins led to astringency perception, but in the presence of other components (ethanol, mannoproteins and glycerol), the intensity of the astringency was significantly lower.

Both trained and expert panels associated ethanol with the perception of 'alcohol feeling', with no significant difference between their scores. As expected, the sample with higher 'alcohol feeling' was WE, containing only the wine and ethanol; followed with a lower score, the sample that contains ethanol plus other ingredients (WEMGT).

The attribute 'cereal taste' scored significantly different between panels for all samples (Table 4a). The trained panel could not differentiate the perception of the 'cereal taste' compared to the other attributes. Whilst the expert panel discriminated the sample according to this attribute, scoring it higher in those samples containing tannins. This finding that oak tannin leads to 'cereal taste' has never been reported. The present authors hypothesised that the 'bitterness' provided by oak tannins could lead panellists to associate it to the 'cereals taste'.

For 'bitterness', both panels agreed that samples containing ethanol and/or tannins are significantly more bitter than the rest (Table 4a), which is in relation to the 'cereal taste' reported previously.

3.2.2. After-Swallowing Sensations

Table 4b shows the after-swallowing sensation scores when consuming the samples. The intensity of all attributes decreased, especially those regarding flavour ('cereal') and taste ('bitterness'), whilst 'body' and 'astringency' values remained similar. There were no significant differences between samples within panels.

It is noteworthy that the attribute 'body' was detectable after swallowing the wine, showing that it is a sensation related, among others, to lingering feelings in-mouth. According to a previous study using tribological techniques [13], samples with glycerol showed higher lubrication. This would indicate that 'body' relates to 'unctuosity'. 'Astringency' is a sensation that also lingers in the mouth, this was not unexpected for the depletion of the mouth's mucous layer caused by polyphenols components [50] and the consequent 'dry' feeling.

3.3. Relationships between Instrumental Measurements and Sensory Analysis

To investigate the relationship between instrumental and sensory measurements, trained panel scores and instrumental correlation analyses were performed (see Table 5). The scores of the trained panel were selected because they discriminated more between samples compared to those of the expert panel. In the same way, as the analysis was used for in-mouth scores, they showed higher values for all attributes' intensity.

Table 5. Pearson's Correlation Coefficient between viscosity at different shear rates and sensory attributes (body perception and astringency).

Shear rate (1 s^{-1}) (Instrumentally Measured)	Body (p-Value)	Astringency (p-Value)
100	0.485 (0.329)	**0.855 (0.030)**
10	0.083 (0.876)	0.582 (0.226)
1	0.032 (0.953)	0.553 (0.255)
0.1	0.118 (0.823)	0.616 (0.192)
K	0.077 (0.884)	0.603 (0.205)
n	0.128 (0.809)	0.692 (0.127)

Values in bold are different from 0 with a significance level alpha = 0.05. Model-wine letters indicate components: base wine (W), ethanol (E), mannoprotein (M), glycerol (G) and tannins (T).

Figure 3 shows the relationship between density and 'body' perception for model-wine samples, with and without saliva. In the absence of saliva (Figure 3a), there was no linear relationship between instrumental and sensory 'body' assessment. For measurements made on samples with added saliva (Figure 3b), which mimic a real consumption scenario, a trend between density and 'body' appeared. Considering only the samples not containing ethanol, there was a positive and significant correlation between instrumental density and 'body' perception ($R = 0.971, p = 0.029$).

A past study demonstrated that for Newtonian fluids, the sensory attribute 'thick' had a high correlation with instrumental viscosity [51]. However, there are no references for non-Newtonian fluids. So, for wine-model samples with saliva, there was a small, non-significant relationship between 'body' and instrumental apparent viscosity in samples with added saliva, at a shear rate of 100 s^{-1}.

Apparent viscosity values were correlated to 'astringency' values ($R = 0.855$, $p = 0.030$) (Figure 3c), confirming that the formation of the protein-polyphenol complex influenced the instrumental viscosity.

Previous work had stated that wine body classification (full bodied, medium bodied and light bodied) was empirical [18–21]. But in this present paper, authors present that the sensory wine attribute 'body' relates to the instrumental density, although not with viscosity.

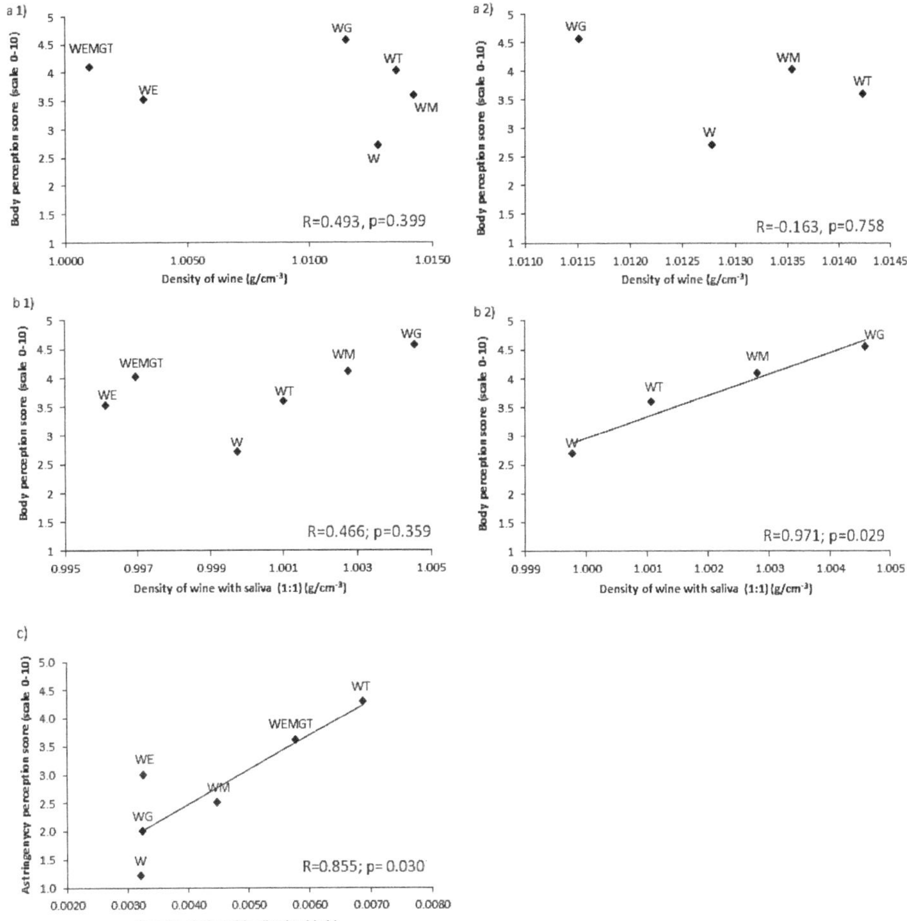

Figure 3. Relationship between instrumental measurement of the model-wine formulations and trained panel scores for: (**a1**) density of samples without saliva, and 'body' scores; (**a2**) density of samples without added ethanol plus saliva, and 'body' scores; (**b1**) instrumental density of samples plus saliva, and 'body' score; (**b2**) instrumental density of samples without ethanol plus saliva, and 'body' score; (c) instrumental viscosity of samples plus saliva and 'astringency score'. Model-wine letters indicate components: base wine (W), ethanol (E), mannoprotein (M), glycerol (G) and tannins (T).

4. Conclusions

The present study shows that glycerol and/or mannoproteins contributed to the 'body' feeling, that was correlated with the instrumental measurement of the density of model-wine samples mixed with saliva. As reported previously, tannins were the leading cause for the 'astringency' feeling and an increase in instrumental viscosity due to the formation of salivary protein-polyphenols complexes. In parallel, this study shows that there is a gap between the abilities of the expert and trained panels for describing wine texture sensations. This makes it even more difficult to understand the expectations and preferences of real consumers regarding wine attributes, such as 'body', which is normally considered the driver for wine quality perception and liking.

Future studies will include real wine samples and saliva for multiple participants. There is a need for further research on how consumers understand the term 'wine body' in commercial samples. Also, determining what are the characteristics of the wine consumers associate with the term 'body' will be beneficial. In addition, knowing how the composition of different wines affects the values of physical properties, such as density or viscosity, will be of immense value.

Author Contributions: Conceptualisation, L.L. and B.B.; Formal analysis, L.L. and E.S.; Investigation, M.D.A.; Methodology, L.L.; Supervision, M.V.M.-A. and B.B.; Writing—original draft, L.L.; Writing—review and editing, L.L., M.D.A., M.V.M.-A. and B.B.

Funding: The authors of this manuscript were funded by the Spanish MINECO projects (AGL2015-64522-C2-R-01 and RTC-2016-4556-1). Laura Laguna would like to thank the Spanish "Juan de la Cierva" program for her contract (ref. FJCI-2014-19907 and IJCI-2016-27427).

Acknowledgments: Authors would like to thank the help of Pilar Jiménez and Yuanyuan Tan. Finally, the authors thank Phillip John Bentley for his help with the English language edition.

Conflicts of Interest: The authors declare no conflict of interest.

References

1. Szczesniak, A.S. Texture is a sensory property. *Food Qual. Prefer.* **2002**, *13*, 215–225. [CrossRef]
2. Bourne, M. *Food Texture and Viscosity: Concept and Measurement*; Academic Press: New York, NY, USA, 2002.
3. DeMiglio, P.; Pickering, G.J.; Reynolds, A.G. Astringent sub-qualities elicited by red wine: The role of ethanol and pH. In Presented at the Proceedings of the International Bacchus to the Future Conference, St Catharines, ON, Canada, 23–25 May 2002; pp. 31–52.
4. Hand, A.R.; Frank, M.E. *Fundamentals of Oral Histology and Physiology*; John Wiley & Sons: Ames, IA, USA, 2014.
5. Jackson, R.S. *Wine Tasting: A Professional Handbook*; Brock University: St Catharines, ON, Canada.
6. Gawel, R.; Waters, E.J. The Effect of Glycerol on the Perceived Viscosity of Dry White Table Wine. *J. Wine Res.* **2008**, *19*, 109–114. [CrossRef]
7. Hopfer, H.; Heymann, H. Judging wine quality: Do we need experts, consumers or trained panelists? *Food Qual. Prefer.* **2014**, *32*, 221–233. [CrossRef]
8. Lawless, H.T.; Heymann, H. *Sensory Evaluation of Food*; Springer: New York, NY, USA, 2010; pp. 1–18.
9. King, E.S.; Dunn, R.L.; Heymann, H. The influence of alcohol on the sensory perception of red wines. *Food Qual. Prefer.* **2013**, *28*, 235–243. [CrossRef]
10. Ferrer-Gallego, R.; Hernández-Hierro, J.M.; Rivas-Gonzalo, J.C.; Escribano-Bailón, M.T. Sensory evaluation of bitterness and astringency sub-qualities of wine phenolic compounds: Synergistic effect and modulation by aromas. *Food Res. Int.* **2014**, *62*, 1100–1107. [CrossRef]
11. Laguna, L.; Sarkar, A. Oral tribology: Update on the relevance to study astringency in wines. *Tribol. Mater. Surf. Interf.* **2017**, *11*, 116–123. [CrossRef]
12. Quijada-Morin, N.; Williams, P.; Rivas-Gonzalo, J.C.; Doco, T.; Escribano-Bailon, M.T. Polyphenolic, polysaccharide and oligosaccharide composition of Tempranillo red wines and their relationship with the perceived astringency. *Food Chem.* **2014**, *154*, 44–51. [CrossRef] [PubMed]
13. Laguna, L.; Sarkar, A.; Bryant, M.G.; Beadling, A.R.; Bartolomé, B.; Victoria Moreno-Arribas, M. Exploring mouthfeel in model wines: Sensory-to-instrumental approaches. *Food Res. Int.* **2017**, *102*, 478–486. [CrossRef]
14. Lubbers, S.; Verret, C.; Voilley, A. The Effect of Glycerol on the Perceived Aroma of a Model Wine and a White Wine. *LWT Food Sci. Technol.* **2001**, *34*, 262–265. [CrossRef]
15. Gil, M.; Quirós, M.; Fort, F.; Morales, P.; Gonzalez, R.; Canals, J.-M.; Zamora, F. Influence of grape maturity and maceration length on polysaccharide composition of Cabernet sauvignon red wines. *Am. J. Enol. Vitic.* **2015**, *66*, 393–397. [CrossRef]
16. Vidal, S.; Francis, L.; Williams, P.; Kwiatkowski, M.; Gawel, R.; Cheynier, V.; Waters, E. The mouth-feel properties of polysaccharides and anthocyanins in a wine like medium. *Food Chem.* **2004**, *85*, 519–525. [CrossRef]
17. Amerine, M.A.; Pangborn, E.B.; Roessler, E.B. Principles of sensory evaluation of food. *Food Sci. Technol. Monogr.* **1965**, 93–101.
18. Gawel, R.; Sluyter, S.V.; Waters, E.J. The effects of ethanol and glycerol on the body and other sensory characteristics of Riesling wines. *Aust. J. Grape Wine Res.* **2007**, *13*, 38–45. [CrossRef]

19. Nurgel, C.; Pickering, G. Contribution of glycerol, ethanol and sugar to the perception of viscosity and density elicited by model white wines. *J. Texture Stud.* **2005**, *36*, 303–323. [CrossRef]
20. Pickering, G.J.; Heatherbell, D.; Vanhanen, L.; Barnes, M. The effect of ethanol concentration on the temporal perception of viscosity and density in white wine. *Am. J. Enol. Vitic.* **1998**, *49*, 306–318.
21. Runnebaum, R.; Boulton, R.; Powell, R.; Heymann, H. Key constituents affecting wine body–an exploratory study. *J. Sensory Stud.* **2011**, *26*, 62–70. [CrossRef]
22. Cliff, M.A.; King, M.C.; Schlosser, J. Anthocyanin, phenolic composition, colour measurement and sensory analysis of BC commercial red wine. *Food Res. Int.* **2007**, *40*, 92–100. [CrossRef]
23. García-Ruiz, A.; Bartolomé, B.; Martínez-Rodríguez, A.J.; Pueyo, E.; Martín-Álvarez, P.J.; Moreno-Arribas, M. Potential of phenolic compounds for controlling lactic acid bacteria growth in wine. *Food Control* **2008**, *19*, 835–841. [CrossRef]
24. Cala, O.; Dufourc, E.J.; Fouquet, E.; Manigand, C.; Laguerre, M.; Pianet, I. The colloidal state of tannins impacts the nature of their interaction with proteins: The case of salivary proline-rich protein/procyanidins binding. *Langmuir* **2012**, *28*, 17410–17418. [CrossRef] [PubMed]
25. Ferrer-Gallego, R.; Bras, N.F.; Garcia-Estevez, I.; Mateus, N.; Rivas-Gonzalo, J.C.; de Freitas, V.; Escribano-Bailon, M.T. Effect of flavonols on wine astringency and their interaction with human saliva. *Food Chem.* **2016**, *209*, 358–364. [CrossRef] [PubMed]
26. Laguna, L.; Bartolomé, B.; Moreno-Arribas, M.V. Mouthfeel perception of wine: Oral physiology, components and instrumental characterization. *Trends Food Sci. Technol.* **2017**, *59*, 49–59. [CrossRef]
27. Taladrid, D.; Lorente, L.; Bartolomé, B.; Moreno-Arribas, M.V.; Laguna, L. An integrative salivary approach regarding palate cleansers in wine tasting. *J. Texture Stud.* **2019**, *50*, 75–82. [CrossRef]
28. Waters, E.J.; Pellerin, P.; Brillouet, J.M. A saccharomyces mannoprotein that protects wine from protein haze. *Carbohydr. Polym.* **1994**, *23*, 185–191. [CrossRef]
29. Pellerin, P.; Doco, T.; Vidal, S.; Williams, P.; Brillouet, J.M.; Oneill, M.A. Structural characterization of red wine rhamnogalacturonan II. *Carbohydr. Res.* **1996**, *290*, 183–197. [CrossRef]
30. Vidal, S.; Courcoux, P.; Francis, L.; Kwiatkowski, M.; Gawel, R.; Williams, P.; Waters, E.; Cheynier, V. Use of an experimental design approach for evaluation of key wine components on mouth-feel perception. *Food Qual. Pref.* **2004**, *15*, 209–217. [CrossRef]
31. Shama, F.; Sherman, P. Identification of stimuli controlling the sensory evaluation of viscosity II. Oral methods. *J. Texture Stud.* **1973**, *4*, 111–118. [CrossRef]
32. Yamagata, Y.; Izumi, A.; Egashira, F.; Miyamoto, K.-I.; Kayashita, J. Determination of a suitable shear rate for thickened liquids easy for the elderly to swallow. *Food Sci. Technol. Res.* **2012**, *18*, 363–369. [CrossRef]
33. Laguna, L.; Farrell, G.; Bryant, M.; Morina, A.; Sarkar, A. Relating rheology and tribology of commercial dairy colloids to sensory perception. *Food Funct.* **2017**, *8*, 563–573. [CrossRef] [PubMed]
34. Neto, F.S.; de Castilhos, M.B.; Telis, V.R.; Telis-Romero, J. Effect of ethanol, dry extract and reducing sugars on density and viscosity of Brazilian red wines. *J. Sci. Food Agric.* **2015**, *95*, 1421–1427. [CrossRef]
35. Santos-Buelga, C.; De Freitas, V. *Wine Chemistry and Biochemistry*; Springer: New York, NY, USA, 2009; pp. 529–570.
36. Moreno-Arribas, V.; Pueyo, E.; Nieto, F.J.; Martin-Alvarez, P.J.; Polo, M.C. Influence of the polysaccharides and the nitrogen compounds on foaming properties of sparkling wines. *Food Chem.* **2000**, *70*, 309–317. [CrossRef]
37. Riou, V.; Vernhet, A.; Doco, T.; Moutounet, M. Aggregation of grape seed tannins in model wine - effect of wine polysaccharides. *Food Hydrocoll.* **2002**, *16*, 17–23. [CrossRef]
38. Haslams, E. *Practical Polypenolics: From Structure to Molecular Recognition and Physiological Action*; Cambridge University Press: Cambridge, UK, 1998.
39. Yanniotis, S.; Kotseridis, G.; Orfanidou, A.; Petraki, A. Effect of ethanol, dry extract and glycerol on the viscosity of wine. *J. Food Eng.* **2007**, *81*, 399–403. [CrossRef]
40. Harrington, R.J. The wine and food pairing process: Using culinary and sensory perspectives. *J. Culin. Sci. Technol.* **2005**, *4*, 101–112. [CrossRef]
41. Niimi, J.; Danner, L.; Li, L.; Bossan, H.; Bastian, S.E. Wine consumers' subjective responses to wine mouthfeel and understanding of wine body. *Food Res. Int.* **2017**, *99*, 115–122. [CrossRef]

42. Pascal, C.; Poncet-Legrand, C.; Cabane, B.; Vernhet, A. Aggregation of a proline-rich protein induced by epigallocatechin gallate and condensed tannins: Effect of protein glycosylation. *J. Agric. Food Chem.* **2008**, *56*, 6724–6732. [CrossRef] [PubMed]
43. Montealegre, R.R.; Peces, R.R.; Vozmediano, J.C.; Gascueña, J.M.; Romero, E.G. Phenolic compounds in skins and seeds of ten grape Vitis vinifera varieties grown in a warm climate. *J. Food Compos. Anal.* **2006**, *19*, 687–693. [CrossRef]
44. Ma, W.; Guo, A.; Zhang, Y.; Wang, H.; Liu, Y.; Li, H. A review on astringency and bitterness perception of tannins in wine. *Trends Food Sci. Technol.* **2014**, *40*, 6–19. [CrossRef]
45. Hufnagel, J.C.; Hofmann, T. Orosensory-directed identification of astringent mouthfeel and bitter-tasting compounds in red wine. *J. Agric. Food Chem.* **2008**, *56*, 1376–1386. [CrossRef] [PubMed]
46. Danilova, V.; Hellekant, G. Oral sensation of ethanol in a primate model III: Responses in the lingual branch of the trigeminal nerve of Macaca mulatta. *Alcohol* **2002**, *26*, 3–16. [CrossRef]
47. Jackson, R.S. *Table Wines: Sensory Characteristics and Sensory Analysis*; Wood Head Publishing: Cambridge, UK, 2012.
48. Meillon, S.; Urbano, C.; Schlich, P. Contribution of the Temporal Dominance of Sensations (TDS) method to the sensory description of subtle differences in partially dealcoholized red wines. *Food Qual. Prefer.* **2009**, *20*, 490–499. [CrossRef]
49. Serafini, M.; Maiani, G.; Ferro-Luzzi, A. Effect of ethanol on red wine tannin– protein (BSA) interactions. *J. Agric. Food Chem.* **1997**, *45*, 3148–3151. [CrossRef]
50. Arnold, G. A tasting procedure for assessing bitterness and astringency. In *Sensory Quality of Food and Beverages: Definition, Measurement, and Control*; Williams, A.A., Atkins, R.K., Eds.; Horwood Ltd.: Chichester, UK, 1983; pp. 109–114.
51. Van Aken, G.A. Modelling texture perception by soft epithelial surfaces. *Soft Matter* **2010**, *6*, 826. [CrossRef]

Article

Compression Test of Soft Food Gels Using a Soft Machine with an Artificial Tongue

Kaoru Kohyama [1,*], Sayaka Ishihara [2], Makoto Nakauma [2] and Takahiro Funami [2]

1 Food Research Institute, National Agriculture and Food Research Organization (NARO), 2-1-12 Kannondai, Tsukuba, Ibaraki 305-8642, Japan
2 San-Ei Gen F. F. I., Inc., 1-1-11 Sanwa-Cho, Toyonaka, Osaka 561-8588, Japan; sayaka-ishihara@saneigenffi.co.jp (S.I.); m-nakauma@saneigenffi.co.jp (M.N.); tfunami@saneigenffi.co.jp (T.F.)
* Correspondence: kaoruk@affrc.go.jp; Tel.: +81-29-838-8054

Received: 25 April 2019; Accepted: 26 May 2019; Published: 29 May 2019

Abstract: Care food is increasingly required in the advanced-aged society. Mechanical properties of such foods must be modified such that the foods are easily broken by the tongue without chewing. When foods are compressed between the tongue and the hard palate, the tongue deforms considerably, and only soft foods are broken. To simulate tongue compression of soft foods, artificial tongues with stiffness similar to that of the human tongue were created using clear soft materials. Model soft gels were prepared using gellan gums. A piece of gel on an artificial tongue was compressed using a texture analyzer. The deformation profile during the compression test was obtained using a video capture system. The soft machine equipped a soft artificial tongue sometimes fractured food gels unlike hard machine, which always fracture gels. The fracture properties measured using the soft machine were better than those obtained from a conventional test between hard plates to mimic natural oral processing in humans. The fracture force on foods measured using this soft machine may prove useful for the evaluation of food texture that can be mashed using the tongue.

Keywords: texture; compression test; artificial tongue; fracture; soft machine; gellan gum gels; care food

---

1. Introduction

Aging has progressed worldwide. The requirement of appropriate care foods for the elderly in our aged society has also increased. Care foods for individuals who have difficulty in mastication must be soft enough to be consumed by compression with the tongue and hard palate without chewing. Japan is the leading country in the context of aging populations, as the number of elderly people aged 65 years and over in Japan is about 34 million, and the percentage of elderly individuals in the population was 27.3% in 2016 [1]. The elderly population will further increase and reach approximately 40 million by 2042. In contrast, the total population has decreased since 2007. Since 2012, the Japanese Ministry of Agriculture, Forestry, and Fisheries has promoted new care foods for people with dysphagia, difficulty in mastication, malnutrition, and future frailty [2]. Smile Care Foods with a red/yellow/blue mark are being promoted to aid consumers in choosing suitable food products in each area of a storefront [2]. One of the categories included in Smile Care Foods (yellow 3) is the tongue-mashable level [2]. However, to date, numerical values for tongue-mashable foods have not been presented. The aim of this study is to determine tongue mashability of soft food gels using a new instrumental test.

The tongue pressure of humans has been widely measured [3–16]. As balloon-type sensors are handy and easy to manage, significant data for various ages, gender, and eating abilities have been accumulated and correlated with other activities [10–12]. The maximum voluntary tongue pressure of healthy adults was reported as 40–100 kPa in these reports, which decreases with age [13]. Subjects with tongue pressure lower than 40 kPa represent reduced tongue strength [14], thus some people may require special care foods [2].

When a soft food is compressed between the tongue and hard palate, the tongue deforms considerably to fracture the food. The tongue tenses during food compression and may become 5–10 times harder than in the relaxed state [16,17]. Elastic modulus is a mechanical parameter as the ratio of force per unit area required for small deformation expressed as strain (ratio of deformation to the initial length) of samples, which indicates stiffness or difficulty in deformation. In our previous study [16], apparent elastic moduli of the human tongue of healthy young subjects were measured. The value was determined at approximately 20% compressive strain. The apparent modulus was 12.2 ± 4.2 kPa and 122.5 ± 58.5 kPa in a relaxed and tense state, respectively. To simulate tongue compression, multiple artificial tongues with apparent modulus similar to that of the human tongue were used to compress soft gels between one of the artificial tongues and metal plate [16,18]. Agar gels with fracture strain of ca. 60% but different fracture forces were broken when the deformation of agar gel was greater than that of the artificial tongue [16]. This suggests that decreased stiffness expressed as decreased modulus or increased deformability of the artificial tongue did not facilitate fracture the food gels. Among several materials tested, an artificial tongue with the apparent modulus of 55 kPa was most suitable to simulate the human behavior. Specifically, softer food gels with fracture stress (force per unit area) less than ca. 50 kPa or a lower fracture force of cylindrical gels (20 mm diameter) of 20 N could be consumed easily using the tongue and palate [16]. When gels with similar fracture forces and different fracture strains were compared, highly deformable gels with greater fracture strain were not broken by the artificial tongue [18]. Although gels having a fracture strain of 70% or over had lower elastic modulus and fracture stress [19], fracture strain became more critical to change the oral processing method from tongue–palate compression to chewing [18].

The artificial tongues were made using silicone rubber in our previous studies [16,18]. The artificial tongue had to be of a size similar to that of the food gel (20 mm in diameter and 10 mm high), and the compression test was stopped at 10 mm after the surface of the gel contacted the upper plate. When gels deformed more than the artificial tongue in the vertical direction, the gels were fractured between the artificial tongue and metal plate [16,18]. However, highly deformable gels were not broken under these conditions [18]. The wider artificial tongue (56 mm in diameter) was suggested to be more suitable to simulate tongue compression of gels with a high fracture strain; however, observation at the fracture point was not possible. The limited test conditions were caused because the silicone rubber was not transparent. The real human tongue is wider than the food gel and may compress the gel more than the gel's initial height (10 mm) during tongue deformation. Observations of deformation of food gels and artificial tongues were difficult as the tongue covered the food gel during compression. Thus, new artificial tongues were prepared with transparent soft materials, and soft gels as model foods as in previous studies [18–21] were compressed between the soft material and hard plate. We tested three urethane gels with varying stiffness (slightly higher, comparable, and lower level) of the human tongue. We report here a pilot study on gel fracture using a soft machine with transparent artificial tongues.

2. Materials and Methods

Model food gels (φ20 × 10 mm) were prepared by mixing low- and high-acylated gellan gums (KELCOGEL™ and KELCOGEL™ LT-100, respectively, San-Ei Gen F. F. I., Inc., Osaka, Japan) as previously described [18–21] with slight modification. The concentrations of gellan gums are shown in Table 1. Sucrose (10% *w*/*w*), a food color (SAN GREEN™ GC-EM, San-Ei Gen F. F. I., 0.2% *w*/*w*) and calcium lactate (0.1% *w*/*w*) were contained in all gels. The food gels were stored in a refrigerator and used within a month after preparation.

Table 1. Formulation of gellan gels used as model soft foods.

Gellan Gel (% *w/w*)	A15	BC15	D15	A20	BC20	D20
Mixture of gellan gum	0.32	0.35	0.30	0.45	0.45	0.43
Low-acylated gellan gum	0.32	0.2625	0.15	0.45	0.3375	0.215
High-acylated gellan gum	0.00	0.0875	0.15	0.00	0.1125	0.215

Artificial tongues were made using urethane transparent gels (HITOHADA^{TM} gel clear type, Exseal Co., Ltd., Mino, Gifu, Japan). Rectangular gels (50 × 50 × 10 mm) and cylindrical gels (φ13 × 10 mm) were prepared, and Asker C hardness was determined as 0, 7, and 15 by the manufacturer. They were named as H0, H7, and H15, hereafter.

Hardness of artificial tongues with 50 × 50 × 10 mm was determined with a durometer (Asker FP type, Kobunshi Keiki Co., Ltd., Kyoto, Japan) at 20 °C. Compression test of the cylindrical urethane gels was conducted using a TA.XT*plus* Texture Analyser with a 50 N load cell (Stable Micro Systems, Surrey, UK). A cylindrical specimen was compressed between a flat alminum plate and stage at a constant rate of 10 mm/s and 20 °C [18–21]. Both the plate and stage were made of aluminum. Compression started from 10 mm above, force was first detected when the upper plate contacted the surface of artificial tongue and increased during compression. The compression stopped at 2.5 mm from the contact. Apparent modulus was determined based on the stress value at 20% compression [18]. Force value at 2 mm deformation (F_2) was read using an Exponent software (ver. 6.1.15.0, Stable Micro Systems). Apparent modulus in kPa was calculated as F_2 (N)/initial sectional area (133 mm^2)/2 (mm) × initial height (measured for each specimen, ca. 10 mm) × 1000.

Fracture properties of food gels were measured under similar conditions using the Texture Analyser. Fracture force and deformation were determined from the first peak of the compression curve as shown in Figure 1a. Fracture strain was obtained as the ratio of the fracture deformation to the initial height of gel.

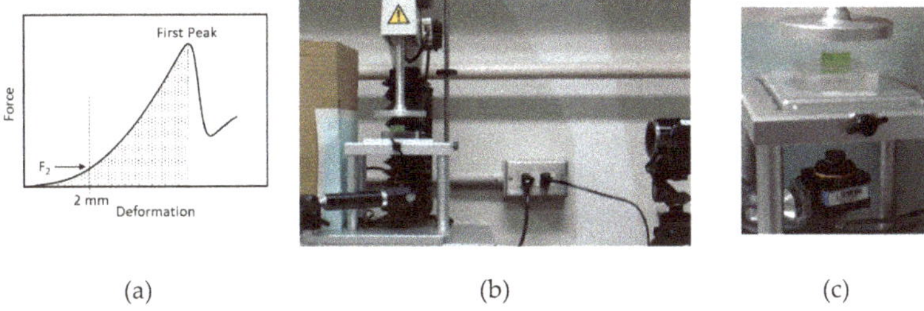

(a) (b) (c)

Figure 1. Typical force–deformation curve of a gellan gel (**a**) and setting for compression test of soft foods between an artificial tongue and plate (**b**,**c**). (**b**) Whole setup and (**c**) enlarged view around the sample. A food gel (φ20 × 10 mm) on a transparent artificial tongue (50 × 50 × 10 mm) was placed on a glass plate of a TA.XT*plus* Texture Analyser. Upper plate connected to a load cell was moved downward at a constant speed, and sample deformation was observed from the bottom and/or from the right side by video cameras during the compression test.

A soft machine was constructed using one of the rectangular artificial tongues and the texture analyzer as shown in Figure 1b. The bottom plate of the Texture Analyser was replaced by a glass plate. A food gel was placed on a rectangle artificial tongue set on the glass plate (Figure 1c) and compressed under similar conditions as above. The F_2, fracture force and fracture deformation were obtained as above. The apparent modulus was calculated assuming the sectional area as 314 mm^2, and fracture work was estimated as shaded area (N.mm) in Figure 1a.

The deformation profile during the compression test was obtained using a Video Capture Synchronisation System (Stable Micro Systems) as in the previous study [19]. Ratios of the cross-sectional area of food gels at fracture point to the initial area were determined by snapshots taken at closest to the fracture point and immediately prior to compression from the bottom video. True fracture stress was calculated as the fracture force divided by the cross-sectional area at the fracture point assuming the initial area of the gels as 314 mm^2 [19]. The compression was stopped when the upper plate (P/75, Stable Micro Systems) contacted the upper surface of the artificial tongue and the load value reached 45 N.

In some cases, deformation of the food gel and artificial tongue was captured from the side using a video camera (Handycam, HDR-XR550V, Sony, Tokyo, Japan) as in the previous reports [16,18] (Figure 1b).

Statistical analyses were performed using a software package (SPSS ver. 23, IBM, Armonk, NY, USA). Statistical significance was set at $p < 0.05$. Mean values for different samples were tested using a one-way analysis of variance (ANOVA) followed by Tukey's multiple comparisons as a *post-hoc* test, as appropriate.

3. Results

3.1. Gellan Gum Gels

Fracture force and strain of food gels measured between metal plates were shown in Table 2. These six gels were prepared to have a low (about 16 N) or high (about 24 N) fracture force, and a low (about 49%), middle (63%) or high (75%), respectively, fracture strain. The nomination is in accordance with a previous study [16].

Table 2. Fracture properties of gellan gels used as soft food models.

Gellan Gel	A15	BC15	D15	A20	BC20	D20
Fracture force (N)	15.7 ^{a} ± 1.1	15.9 ^{a} ± 0.6	16.8 ^{a} ± 1.2	22.5 ^{b} ± 0.9	24.2 ^{bc} ± 2.1	25.0 ^{c} ± 1.5
Fracture strain (%)	49.0 ^{a} ± 2.3	62.0 ^{b} ± 0.4	74.4 ^{c} ± 1.5	49.3 ^{a} ± 2.1	64.1 ^{b} ± 2.4	75.8 ^{c} ± 1.0

Determined at 20 °C 1 day after preparation. Mean and standard deviation values of 6 samples. Values with the same alphabetical letter within a row are not significantly different ($p > 0.05$).

3.2. Artificial Tongues

The mechanical properties of the artificial tongues are shown in Table 3. As these gels did not fracture, the apparent modulus values determined at 20% compression were given. The artificial gels were stable after the compression test and recovered, even after food gels or the of texture analyzer probe (φ20) were inserted. The apparent modulus at 20 °C was measured repeatedly to test long-time stability at room temperature. As shown in Figure 2, they were similar for over a year, and variance values were small. According to the manufacturer, heat tolerance of the urethane gels is between −30 and 80 °C. Therefore, the test could be conducted at body temperature.

Table 3. Characteristics of urethane gels used as artificial tongues.

Gel Sample	H0	H7	H15
Asker FP hardness	43.8 ^{a} ± 1.5	69.7 ^{b} ± 0.6	89.5 ^{c} ± 0.6
Apparent modulus (kPa)	23.3 ^{a} ± 0.3	55.1 ^{b} ± 0.9	160.9 ^{c} ± 2.2

Mean and standard deviation values of 4 or more samples. Values with different alphabetical letter within a row are significantly different ($p < 0.05$).

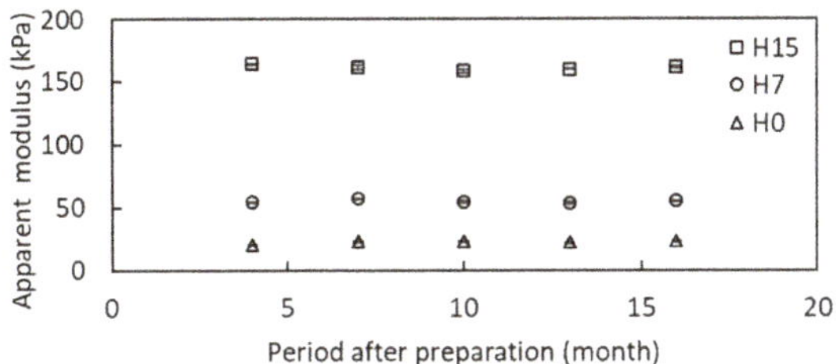

Figure 2. Apparent modulus of urethane gel for artificial tongue. Error bars represent standard errors of 2–7 samples.

3.3. Compression of Gellan Gum Gels Placed on An Artificial Tongue

Gellan gels on an artificial tongue were compressed using a soft machine as shown in Figure 1. When softest artificial tongue H0 was used, harder gellan gels (A20, BC20, and D20) were not broken, while softer gels with 16 N fracture force sometimes fractured (Table 4). Fracture probability decreased as the fracture strain increased (A15 > BC15 > D15). H7 and H15 fractured all food gels as shown in Table 4.

Table 4. Ratio of food gel-fracture on artificial tongue.

Gel Sample	H0	H7	H15
A15	0.83	1.00	1.00
BC15	0.33	1.00	1.00
D15	0.29	1.00	1.00
A20	0.00	1.00	1.00
BC20	0.00	1.00	1.00
D20	0.00	1.00	1.00

Probability from 3 or more samples.

Figure 3 shows compression curves of a food gel (BC20) on different artificial tongues. The force values at a given deformation before fracture decreased as stiffness of the artificial tongues decreased. A soft machine equipped with an artificial tongue H15 showed a sharp peak at fracture as observed in the compression test using a hard machine. The fracture point became less sharp around 9 mm for the middle artificial tongue H7. The food gel never fractured using the softest H0 as shown in Table 4.

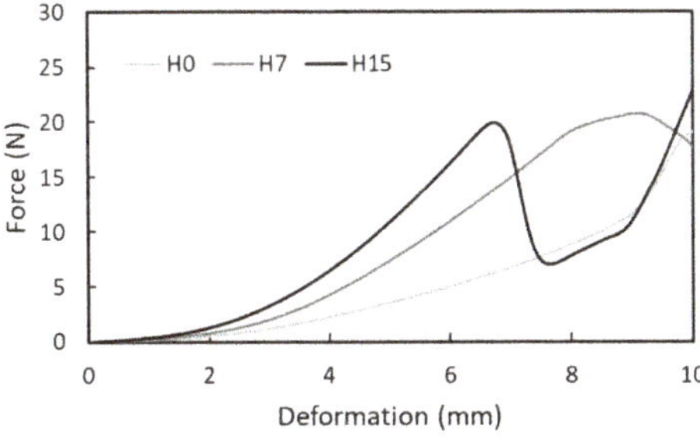

Figure 3. Examples of compression curves of food gel (BC20) on artificial tongues.

When food gels were compressed with the soft machine, phenomena different from those for the conventional test using a hard machine were observed. The transparent material was useful for direct observation during the compression test. Snapshots were taken from the video showing side and bottom views of the samples (Figure 4). During the first 2 mm compression as presented in side views at 2 mm in Figure 4, a food gel was on the artificial tongue. Further compression pushed and inserted the food gel into the artificial tongue. The insertion degree was greater for softer artificial tongues that are presented in side views at 5 mm (H15 < H7 < H0) and at 8 mm (H7 < H0) in Figure 4. If a food gel did not fracture, it was surrounded by artificial tongue when the distance became 10 mm, which was the initial thickness of the artificial tongue (Figure 4c). As the load was applied to a wide surface of the artificial tongue ("bottom" pictures of Figure 4c, the food gel was protected by the soft material from further deformation. When the upper probe was removed after the compression test, the food gel was not fractured, although some syneresis was observed as shown in the "end" picture of Figure 4c. Further, the artificial tongue recovered to its initial state (end pictures of Figure 4c). This condition is different from compression between an artificial tongue with the same-sized food gel and hard plate [16,18], in which the food gel deforms less on the wider soft material. This is the reason why the A15 gel with the lowest fracture strain exhibited a higher fracture probability (Table 3).

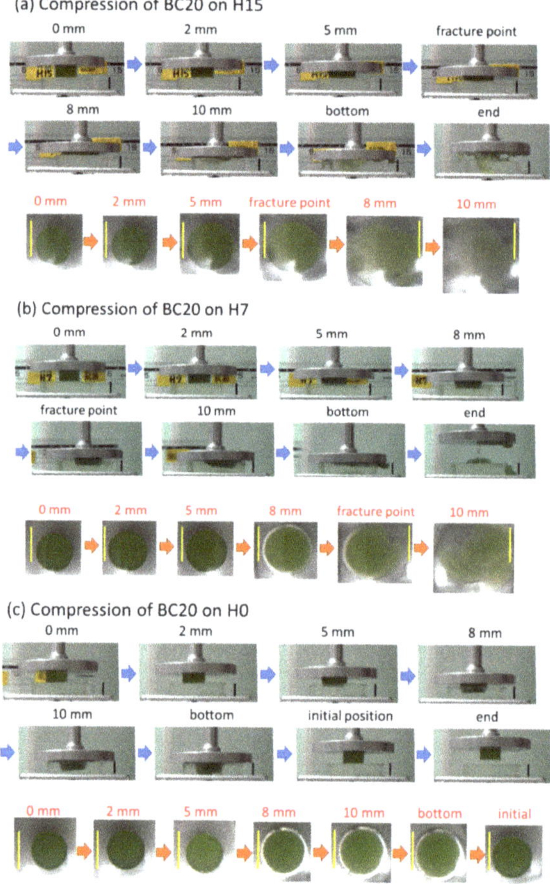

Figure 4. Examples of a food gel on soft material (i.e. as artificial tongue) observed from the side and bottom. Compression process of a BC gel specimen on (**a**) H15, (**b**) H7, and (**c**) H0 artificial tongue. Compression distance from the first contact to the food gel is shown over each snapshot. "Bottom" represents the moment that compression was stopped due to the limit force, "initial position" is the moment that the probe returned to 0 mm height, and "end" indicates the test completion. Black bars are 10 mm as original height of food and artificial tongue gels in the side views, and yellow bars are 20 mm as original diameter of food gels in the bottom views.

3.4. Mechanical Properties of Gellan Gum Gels Measured Using the Soft Machine

Table 5 shows the apparent moduli and some fracture properties measured with the soft machine. To calculate the apparent moluli, we assumed food gel deformation at only 2 mm compression as video observation from the side revealed that deformation of the artificial tongues was not significant. As shown in Figure 4, the bottom area was not significantly expanded at 2 mm compression; therefore, the initial sectional area (314 mm^2) was used to calculate the apparent moluli.

Table 5. Mechanical characteristics of gellan gels measured by the soft machine.

Gellan Gel	A15	BC15	D15	A20	BC20	D20
on H15 artificial tongue						
Apparent modulus (kPa)	$13.4^{ab} \pm 4.7$	$13.2^{abB} \pm 0.3$	$2.6^{a} \pm 0.9$	$32.7^{c} \pm 12.2$	$18.5^{bcB} \pm 1.2$	$5.0^{abA} \pm 0.8$
Fracture force (N)	$10.9^{aA} \pm 3.5$	$12.7^{ab} \pm 1.4$	$12.5^{a} \pm 0.3$	$17.2^{abc} \pm 3.5$	$19.3^{bc} \pm 0.5$	$22.6^{c} \pm 2.8$
Fracture deformation (mm)	$5.95^{aA} \pm 0.83$	$6.53^{abA} \pm 0.13$	$7.75^{cdA} \pm 0.15$	$5.55^{abA} \pm 0.48$	$6.89^{bcA} \pm 0.19$	$8.17^{dA} \pm 0.38$
Fracture work (N.mm)	$21.8^{aA} \pm 3.3$	$27.5^{a} \pm 1.3$	$23.0^{a} \pm 1.6$	$32.0^{bA} \pm 9.1$	$42.1^{bcA} \pm 1.2$	$44.8^{c} \pm 5.4$
Area ratio at fracture to initial	$1.49^{abB} \pm 0.10$	$1.76^{ab} \pm 0.10$	$2.00^{ab} \pm 0.70$	$1.42^{a} \pm 0.13$	$1.75^{ab} \pm 0.31$	$2.42^{b} \pm 0.11$
True fracture stress (kPa)	$27.1^{aA} \pm 5.9$	$23.0^{a} \pm 2.0$	$22.1^{a} \pm 9.7$	$38.3^{aA} \pm 5.3$	$35.9^{a} \pm 6.3$	$29.6^{a} \pm 3.3$
on H7 artificial tongue						
Apparent modulus (kPa)	$23.1^{c} \pm 3.4$	$5.4^{aA} \pm 4.6$	$2.3^{a} \pm 1.0$	$20.0^{bc} \pm 10.4$	$8.3^{abA} \pm 1.5$	$2.6^{aA} \pm 1.4$
Fracture force (N)	$13.0^{a} \pm 1.7$	$12.9^{a} \pm 4.5$	$14.6^{ab} \pm 3.4$	$21.7^{ab} \pm 2.1$	$20.8^{ab} \pm 5.7$	$23.7^{b} \pm 5.5$
Fracture deformation (mm)	$7.00^{aA} \pm 0.27$	$7.78^{abB} \pm 0.49$	$9.16^{cB} \pm 0.85$	$8.82^{bcB} \pm 0.46$	$9.41^{cB} \pm 0.32$	$9.93^{cB} \pm 0.37$
Fracture work (N.mm)	$35.5^{abB} \pm 5.4$	$29.9^{a} \pm 13.4$	$34.0^{ab} \pm 10.5$	$71.6^{cB} \pm 5.7$	$68.4^{cB} \pm 15.7$	$61.6^{bc} \pm 10.5$
Area ratio at fracture to initial	$1.42^{aAB} \pm 0.04$	$1.60^{ab} \pm 0.36$	$2.00^{b} \pm 0.16$	$1.33^{a} \pm 0.06$	$1.50^{a} \pm 0.05$	$2.59^{c} \pm 0.14$
True fracture stress (kPa)	$29.2^{abA} \pm 3.9$	$25.0^{a} \pm 5.0$	$23.1^{a} \pm 4.6$	$52.3^{cB} \pm 5.6$	$43.8^{bc} \pm 10.8$	$29.3^{ab} \pm 8.2$
on H0 artificial tongue; part of A15, BC15, D15 gels fractured, A20, BC20 and D20 not fractured						
Apparent modulus (kPa)	$11.3^{a} \pm 6.9$	$7.0^{aAB} \pm 2.6$	$2.4^{a} \pm 1.6$	$10.3^{a} \pm 3.3$	$9.9^{aA} \pm 3.0$	$4.4^{aA} \pm 1.1$
Fracture force (N)	$19.8^{B} \pm 1.39$	-	-	NF	NF	NF
Fracture deformation (mm)	$10.3^{B} \pm 0.39$	-	-	NF	NF	NF
Fracture work (N.mm)	$56.1^{C} \pm 6.7$	-	-	NF	NF	NF
Area ratio at fracture to initial	$1.29^{A} \pm 0.05$	-	-	NF	NF	NF
True fracture stress (kPa)	$49.2^{B} \pm 4.1$	-	-	NF	NF	NF

Mean ± standard deviation of at least three samples. For H0 artificial tongue, fracture properties are presented for only A15 gels with high fracture probability (0.83). Mechanical values with the same lower-case letter within a row are not significantly different among gellan samples, and those with the same upper-case letter within a column are not significantly different among artificial tongues ($p > 0.05$). Row or column without superscripts are not significantly different by one-way ANOVA ($p > 0.05$).

The fracture forces were similar among A15, BC15, and D15, and a somewhat higher fracture force was observed in D20 than in A20 and BC20. Fracture deformation increased in the order A < BC < D for both groups. Those tendencies were similar to results in the conventional compression test (Table 2). The area ratio at fracture increased as fracture deformation increased. Further, the true stress tended to be higher in 20 series than in 15 series and decreases in the order A > BC > D.

When comparing the effect of different artificial tongues H15 and H7 for the same food gels, the fracture deformation with stiffer H15 was shorter than that with softer H7. Changes in fracture force and true fracture stress were small. Fracture work increased due to fracture deformation as the stiffness of the artificial tongue decreased. Further compression of food gel being surrounded with the artificial tongue was impossible, thus deformation mainly determined gel fracturing.

Using the softest material of H0, series of gels with original fracture force were higher than 20 N, and most BC15 and D15 with original fracture force of 16 N gels were not broken. Even for fractured gels, no clear peak at the fracture point was observed. Thus, Table 5 includes fracture data of A15 on H0 data. The fracture deformation of A15 gels that fractured at a high probability (0.83) was calculated as 10.3 mm. It suggests that the gel was inserted into the artificial tongue without fracture as shown in the side view of Figure 3. Some of D15, BC20, and D20 on H7 are suspected to have similar fractures. These cases with 10 mm or higher fracture deformation may be fractured during decompression.

4. Discussion

In this study, six mouthful-sized food gels ($\varphi 20 \times 10$ mm) were prepared with gellan gum mixtures as shown in Table 2. Previous results on gellan gels of the same size revealed that gels with a 15 N fracture force were easily fractured between the tongue and hard palate by healthy young adults [18]. When the fracture force was 20 N or higher, some gels were not broken between the tongue and palate. We adjusted the fracture force of gellan gels to about 16 N and slightly higher than 20 N. As expected, they could easily fracture and hard gels by compression between the tongue and palate. A small or large fracture strain was seen in the series A or D gels in our previous study [16], and the gels were similar to A15 and A20, or D15 and D20. As gels with the middle fracture strain were between the previous series of B and C [18], they were named BC15 and BC20. As shown in Table 2, D20 had significantly higher fracture force than A20. The other points were suitable as expected.

The maximum isometric tongue pressure of humans has been reported as in the order of 10 kPa [3–12]. We used a 50 N load cell, as it was most suitable to measure force applied on the food gels that fractured at 15–25 N (Table 2). The sectional area of food gels was first 314 mm^2 and may expand 1.5–3.0 times during compression until fracture [19]. The artificial tongues could not fracture by the force of 50 N, the soft machine is suitable to simulate compression of soft foods between the tongue and palate.

The compression test using the soft material fractured some soft gels and did not fracture difficult type gels with tongue–palate compression. A conventional test between hard plates fracture all food gels. The soft machine was better than conventional hard machine to mimic natural oral processing of humans. The behavior of food gels between an artificial tongue and hard plate is useful for evaluation of food texture that can be mashed using the tongue. The transparent materials were stable for practical use and could demonstrate fracture or non-fracture process of food gels between the tongue and hard palate.

True fracture stress of gels having similar fracture force values tended to decrease with increasing fracture strain of original gels as deformation increased. When food gels were compressed with a soft machine, soft material deformed considerably and protected food gels from further deformation. As stiffness or apparent modulus of the artificial tongue decreased, fracture point of gels was increased. However, gel deformation in the horizontal direction was limited in the soft material. Fracture of food gels did not occur in the combination of the softest artificial tongue and food gels with the high fracture force (>20 N). Fracture force of 16 N appeared at the fracture border with the artificial tongue, suggesting an increase in the fracture of food gels with the fracture strain.

The hardest artificial tongue (H15) with a wide sectional area of 2500 mm^2 did not deform until reaching 2 mm (20% strain) under 50 N force. The apparent modulus of the artificial tongues must be measured using small cylindrical gels using the load cell. The apparent modulus is useful parameter to compare mechanical properties of different artificial tongues, the human tongue, and food gels.

Apparent modulus of the human tongue is 12–123 kPa [16] and is in accordance with other results [3–13,15]. We tested three urethane gels. The hardest H15 with the apparent modulus of 161 kPa was too hard to simulate natural eating behavior using the tongue. This may be useful for an artificial gingival rather than a tongue model. As one level higher Smile Care Foods (yellow 4) are slightly harder food that can be mashed with the gums [2].

The middle H7 had an apparent modulus of 55 kPa that was just in the range of the human tongue. The apparent modulus was similar to that of a previously tested silicone rubber A5, which was 53.5 kPa [16,18]. All food gels tested were fractured with H7, as suggested in a previous report [18]. Although the artificial tongue can fracture these gels when the size is greater than the food gels, the fracture probabilities were low if the food gels and artificial tongue were of similar size [18]. It seemed too hard to mimic tongue compression, especially for the elderly individuals who require nursing care foods.

The softest H0 has an apparent modulus of 23 kPa. This value was lower than the maximum tongue pressure in healthy subjects and similar to that in frail elderly subjects [12] and in majority of elder people aged over 80 years [11]. It may be a suitable model tongue for persons with weak tongue pressure who require nursing care foods. H0 may be too soft as a tongue model of healthy adults because the human tongue tenses to mush food in the oral cavity. Because soft gels with fracture force of 15 N were on the fracture border, it is reasonable that the human tongue normally tenses to mash food in the oral cavity. A new artificial tongue that has mechanical properties closer to those of the human tongue in the tensed state is required to simulate the human behavior more precisely.

5. Conclusions

Soft food gels were compressed using three transparent urethane gels. The soft machine could demonstrate gel fracture between the tongue and hard palate of humans. The fracture properties obtained from the test could be useful to evaluate food texture that can be mashed using the tongue.

Author Contributions: K.K. conceived and designed the experiments; K.K. and S.I. performed the experiments and analyzed the data; K.K., S.I. and M.N. contributed materials; K.K. wrote original draft; K.K., S.I., M.N. and T.F. reviewed and edited the paper.

Funding: Part of this study was supported by JSPS Grant-in-Aid for Scientific Research (B) 19H01618 for K.K.

Acknowledgments: English language was edited by Editage.

Conflicts of Interest: The authors declare no conflict of interest.

References

1. Annual Report on the Aging Society: 2017 (Summary). Available online: https://www8.cao.go.jp/kourei/english/annualreport/2017/2017pdf_e.html (accessed on 28 February 2019).
2. Efforts relating to "Smile Care Food". Available online: http://www.maff.go.jp/e/policies/food_ind/attach/pdf/index-9.pdf (accessed on 28 February 2019).
3. Ono, T.; Hori, K.; Masuda, Y.; Hayashi, T. Recent advances in sensing oropharyngeal swallowing function in Japan. *Sensors* **2010**, *10*, 176–202. [CrossRef] [PubMed]
4. Clark, H.M.; Henson, P.A.; Barber, W.D.; Stierwalt, J.A.G.; Sherrill, M. Relationships among subjective and objective measures of tongue strength and oral phase swallowing impairments. *Am. J. Speech Lang. Pathol.* **2003**, *12*, 40–50. [CrossRef]
5. Youmans, S.R.; Youmans, G.L.; Stierwalt, J.A. Differences in tongue strength across age and gender: Is there a diminished strength reserve? *Dysphagia* **2009**, *24*, 57–65. [CrossRef] [PubMed]
6. Alsanei, W.A.; Chen, J.; Ding, R. Food oral breaking and the determining role of tongue muscle strength. *Food Res. Int.* **2015**, *67*, 331–337. [CrossRef]

7. Magalhães, H.V., Jr.; Tavares, J.C.; Magalhães, A.A.B.; Galvão, H.C.; Ferreira, M.A.F. Characterization of tongue pressure in the elderly. *Audiol. Commun. Res.* **2014**, *19*, 375–379. [CrossRef]
8. Jeong, D.-M.; Shin, Y.-J.; Lee, N.-R.; Lim, H.-K.; Choung, H.-W.; Pang, K.-M.; Kim, B.-J.; Kim, S.-M.; Lee, J.-H. Maximal strength and endurance scores of the tongue, lip, and cheek in healthy, normal Koreans. *J. Korean Assoc. Oral Maxillofac. Surg.* **2017**, *43*, 221–228. [CrossRef] [PubMed]
9. van Ravenhorst-Bell, H.A.; Mefferd, A.S.; Coufal, K.L.; Scudder, R.; Patterson, J. Tongue strength and endurance: Comparison in active and non-active young and older adults. *Int. J. Speech Lang. Pathol.* **2017**, *19*, 77–86. [CrossRef]
10. Sakai, K.; Nakayama, E.; Tohara, H.; Maeda, T.; Sugimoto, M.; Takehisa, T.; Takehisa, Y.; Ueda, K. Tongue strength is associated with grip strength and nutritional status in older adult inpatients of a rehabilitation hospital. *Dysphagia* **2017**, *32*, 241–249. [CrossRef] [PubMed]
11. Nagayoshi, M.; Higashi, M.; Takamura, N.; Tamai, M.; Koyamatsu, J.; Yamanashi, H.; Kadota, K.; Sato, S.; Kawashiri, S.; Koyama, Z.; et al. Social networks, leisure activities and maximum tongue pressure: Cross-sectional associations in the Nagasaki Islands Study. *BMJ Open* **2017**, *7*, e014878. [CrossRef] [PubMed]
12. Yamanashi, H.; Shimizu, Y.; Higashi, M.; Koyamatsu, J.; Sato, S.; Nagayoshi, M.; Kadota, K.; Kawashiri, S.; Tamai, M.; Takamura, N.; et al. Validity of maximum isometric tongue pressure as a screening test for physical frailty: Cross-sectional study of Japanese community-dwelling older adults. *Geriatr. Gerontol. Int.* **2018**, *18*, 240–249. [CrossRef] [PubMed]
13. Kikutani, T.; Tamura, F.; Nishiwaki, K.; Kodama, M.; Suda, M.; Fukui, T.; Takahashi, N.; Yoshida, M.; Akagawa, Y.; Kimura, M. Oral motor function and masticatory performance in the community-dwelling elderly. *Odontology* **2009**, *97*, 38–42. [CrossRef] [PubMed]
14. Steele, S.M.; Cichero, J.A.Y. Physiological factors related to aspiration risk: A systematic review. *Dysphagia* **2014**, *29*, 295–304. [CrossRef] [PubMed]
15. Utanohara, Y.; Hayashi, R.; Yoshikawa, M.; Yoshida, M.; Tsuga, K.; Akagawa, Y. Standard values of maximum tongue pressure taken using newly developed disposable tongue pressure measurement devise. *Dysphagia* **2008**, *23*, 286–290. [CrossRef] [PubMed]
16. Ishihara, S.; Nakao, S.; Nakauma, M.; Funami, T.; Hori, K.; Ono, T.; Kohyama, K.; Nishinari, K. Compression test of food gels on artificial tongue and its comparison with human test. *J. Texture Stud.* **2013**, *44*, 104–114. [CrossRef]
17. Shibata, A.; Higashimori, M.; Ramirez-Alpizar, I.G.; Kaneko, M. Tongue elasticity sensing with muscle contraction monitoring. 2012. Tongue elasticity sensing with muscle contraction monitoring. In Proceedings of the 2012 ICME International Conference on Complex Medical Engineering, Kobe, Japan, 1–4 July 2012; pp. 511–516.
18. Ishihara, S.; Isono, M.; Nakao, S.; Nakauma, M.; Funami, T.; Hori, K.; Ono, T.; Kohyama, K.; Nishinari, K. Instrumental uniaxial compression test of gellan gels of various mechanical properties using artificial tongue and its comparison with human oral strategy for the first size reduction. *J. Texture Stud.* **2014**, *45*, 354–366. [CrossRef]
19. Kohyama, K.; Gao, Z.; Watanabe, T.; Ishihara, S.; Nakao, S.; Funami, T. Relationships between mechanical properties obtained from compression test and electromyography variables during natural oral processing of gellan gum gels. *J. Texture Stud.* **2017**, *48*, 66–75. [CrossRef]
20. Gao, Z.; Nakao, S.; Ishihara, S.; Funami, T.; Kohyama, K. A pilot study on ultrasound elastography for evaluation of mechanical characteristics and oral strategy of gels. *J. Texture Stud.* **2016**, *47*, 152–160. [CrossRef]
21. Gao, Z.; Ishihara, S.; Nakao, S.; Hayakawa, F.; Funami, T.; Kohyama, K. Texture evaluation of soft gels with different fracture strains using an artificial tongue. *J. Texture Stud.* **2016**, *47*, 496–503. [CrossRef]

Article

Effect of Oral Physiology Parameters on In-Mouth Aroma Compound Release Using Lipoprotein Matrices: An In Vitro Approach

Amparo Tarrega [1,†], Claude Yven [1,‡], Etienne Semon [1,2], Patrick Mielle [1] and Christian Salles [1,*]

1 CSGA (Centre des Sciences du Goût et de l'Alimentation), AgroSup Dijon, CNRS, INRA, Université de Bourgogne Franche-Comté, F-21000 Dijon, France; atarrega@iata.csic.es (A.T.); Claude.YVEN@agencerecherche.fr (C.Y.); etienne.semon@orange.fr (E.S.); patrick.mielle.inra@free.fr (P.M.)

2 ChemoSens Platform, CSGA, F-21000 Dijon, France

* Correspondence: christian.salles@inra.fr; Tel.: +33-380-69-30-79

† Present address: Instituto de Agroquímica y Tecnología de Alimentos (IATA-CSIC), Paterna, 46980 Valencia, Spain.

‡ Present address: French National Research Agency (ANR), F-75012 Paris, France.

Received: 17 January 2019; Accepted: 15 March 2019; Published: 21 March 2019

Abstract: Temporal aroma compound release during eating is a function of the physicochemical properties of the food matrix, aroma compounds, and oral physiology of individuals. However, the influence of each parameter on the release of each aroma component should be clarified. Two flavored lipoprotein matrices varying in composition were chewed in a chewing simulator that reproduced most of the physiological functions of the mouth. Aroma compound releases (butanoic acid, 2-heptanone, ethyl butyrate, 3-octanone, and 2-nonanone) were followed in real time by direct connection of the device to APCI-MS (atmospheric pressure chemical ionization mass spectrometry). Each oral parameter was controlled and decoupled using the in vitro device. The food matrix composition had only a low impact on aroma compound release, but the controlled oral parameters had significantly different influences on the release of aroma compounds according to their physicochemical characteristics. The release of certain compounds seemed more sensitive to bite force, while others seemed more sensitive to the shearing angle. The salivary flow rate primarily influenced the more hydrophobic compounds. Significant interactions were also observed between shear angle, salivary flow rate, and lipoprotein matrix composition, mainly for the release of the more hydrophobic volatile compounds; this needs further investigations to be clarified.

Keywords: lipoprotein matrix; chewing simulator; aroma compound; in vitro; oral parameters; flavor release

---

1. Introduction

Food oral processing is a complex phenomenon involving several physical, chemical, and physiological mouth functions, leading to food breakdown and changes in in-mouth flavor release and perception before swallowing [1]. The main consequences of the oral process are the food matrix breakdown by mastication and impregnation of food by saliva, which are subjected to several physical, chemical, and biochemical operations [1,2]. During this step of food oral processing, the perception of flavor and texture are progressively affected. The release of aroma and taste compounds from food and their delivery to receptors are key factors leading to flavor perception. The breakdown of food into particles under eating conditions increases the surface area of the food that is exposed to saliva and air; this process drives the release of volatile compounds into the gas phase of the oro-nasal cavity. Moreover, aroma compound release is influenced by several physicochemical factors. During chewing and swallowing, aroma compound release, currently evaluated by nosespace analysis using online mass spectrometry [3], is dependent on food composition and texture, saliva and mastication parameters, as well as the nature of the aroma compounds in terms of partition coefficient and lipophilia [4–6]. In fat-containing foods, the retention and release of aroma compounds mainly depends on their solubility in the aqueous and oil phases and thus on their hydrophobicity [7,8], which is expressed by the log P value [9]. Most of the aroma compounds are more soluble in fat than in water. They are considered as hydrophobic (log P > 1) and less released in the gas phase from oil than from water, while the hydrophilic aroma compounds (log P < 1) are less retained by the oil phase and thus present quite similar air/water and air/oil partition coefficients. Consequently, the release of aroma compounds in the mouth from low-fat foods is higher than that with regular fat foods [6,10,11]. Flavor release and perception are mainly consequences of oral processing. In this process, saliva has various effects on aroma compounds, such as dilution, molecular interactions with saliva components, and enzymatic conversion. It also participates in aroma release. Consequently, it is a major actor in the perception of aroma [12]. A large inter-individual variability has been reported for salivary composition, in particular for alpha-amylase [13,14] and mucins [15]. The variability in composition may be related to variability in perception [16–18].

During this process, aroma compound release varies according to oral parameters [19] that are very dependent on individual oral physiology [5,20,21]. The release of flavor compounds from the food and their delivery to receptors are key factors leading to flavor perception. Temporal in vivo studies currently enable the direct release-perception tests because it is the most natural and realistic way to study flavor release and perception and to take into account all the complexity of the oral process involved in flavor release. However, such in vivo analyses may encounter limitations, such as major inter-individual differences due mainly to physiological and anatomical differences and moderate intra-individual reproducibility. In addition, the food sample must be acceptable to panelists, and only a limited number of samples can be evaluated daily. Another major limitation is the complexity of the oral functions and their interactions. The main consequence is difficulty for the specific role of each oral and product parameter to fully explain flavor release and perception. Thus, artificial devices have been developed to mimic human flavor release during food consumption under controlled conditions and to overcome these in vivo limitations [22–24]. It is a very hard task to simulate and integrate all the oral functions in a single and universal system. Thus, specificities and functionalities of each system should be conceived and tailored to address the asked scientific questions.

Several devices have been developed to simulate human flavor release and bolus formation under mouth-like conditions [23]. Many of them are limited to a controlled stirring or shearing of food particles in a liquid, while volatile compound released in the headspace is sampled [25]. More recently, devices were developed with higher functionality levels. Although they are far from mimicking all the main human mouth functions, mechanical food breakdown was more realistically reproduced, leading to relevant flavor release kinetics produced under controlled conditions. As examples, a mechanical chewing device was specifically developed to study the dynamic release of volatile compounds by direct connection to a proton transfer reaction mass spectrometer (PTR-MS) and sweeteners by saliva sampling analysis from chewing gums [26], although this system reproduced real human mastication patterns rather poorly. Additionally, an innovative and dynamic model mouth was specifically developed to investigate whether the intraoral pressure produced by the tongue affects the release of odorant compounds [27] by connecting online to a PTR-MS apparatus.

A chewing simulator allowing fluid sampling that integrates most of the main human oral functions has been described previously [24,28]. In particular, the shear force, compression force, salivary flow rate, and tongue function reflect average realistic physiological parameters obtained from human subjects and can be decoupled. In this study, we aimed to explore the variations in in vitro aroma release on model cheeses when oral parameters were decoupled using this chewing simulator [24]. The main goals were to test the suitability of the chewing simulator to study differences in flavor release according to parameters based on human oral physiology and to tentatively elucidate the influence of each oral parameter on the volatile compound kinetics, the nature of these compounds, and the composition of the model cheese.

2. Materials and Methods

2.1. Materials

2.1.1. Chemicals

The volatile compounds used—butanoic acid, 2,3-butanedione, 2-heptanone, 2-nonanone, 3-octanone, ethyl hexanoate, ethyl butanoate, dimethyldisulphide—were food grade (Aldrich, St Quentin Fallavier, France). Pure water was prepared using a Milli-Q system (Millipore, St Quentin en Yvelines, France). Anhydrous milk fat (MF-Cormans, Goe-Limbourg, Belgique), powder of milk proteins (dry matter > 950 g/kg; free of fat) (Eurial Poitouraine, Nantes, France), sodium chloride (Jerafrance, Jeufosse, France), and rennet (Labo ABIA, Meursault, France) were used to prepare the lipoprotein matrices. All ingredients were food grade and assessed for microbiological safety.

2.1.2. Lipoprotein Matrix

- Preparation

Flavored lipoprotein matrices (LPMs) were generated by the action of rennet on a mixture of milk protein, milk fat, NaCl, and water [29], to which the aroma compounds were added. Two LPMs were prepared, differing in their fat/milk powder ratio (Table 1): 0.5 for LPM1 and 1 for LPM2. The aroma compound solution was made of (in mg/kg): butanoic acid (10), 2,3-butanedione (3), 2-heptanone (5), 2-nonanone (5), 3-octanone (5), ethyl hexanoate (5), ethyl butanoate (4), and dimethyldisulphide (10) diluted in polyethyleneglycol (1 mL). Rennet extract containing 520 mg/l of chymosin (Sanofi Bio-Industries, Paris, France) diluted in 9 volumes of water was used for the coagulation step. This aroma formulation was optimized in the laboratory from internal cheese aroma analyses data to provide a standard cheese note that was acceptable by the consumers.

Table 1. Composition and characteristics of the lipoprotein matrix (LPM) samples (% of total mass).

	LPM 1	LPM 2
Composition for 500 g		
Water (g)	310 (62)	275 (55)
Anhydrous milk fat (g)	61.5 (12.3)	110 (22)
Milk powder (g)	123.5 (24.7)	110 (22)
NaCl (g)	5 (1)	5 (1)
Aromatic solution [a] (mL)	0.5 (0.1)	0.5 (0.1)
Rennet [b] (mL)	4.8 (0.96)	4.8 (0.96)
Fat/Milk Protein powder	0.5	1
Rheological characteristics		
MD (kPa)	32.78 *	44.38 *
Df (-)	0.42	0.40
Cf (kPa)	23.36	23.88
Wf (kJ/m^3)	4.14	4.34

* Significantly different ($F = 7.76$; $p = 0.022$); [a] Composition: butanoic acid (10 µL), 2,3-butanedione (3 µL), 2-heptanone (5 µL), 2-nonanone (5 µL), 3-octanone (5 µL), ethyl hexanoate (5 µL), ethyl butanoate (4 µL), dimethyl disulphide (10 µL), and polyethyleneglycol up to 1 mL; [b] active chymosin 520 mg/L, dilution 1/10; MD: modulus of deformability, Df: fracture strain, Cf: fracture stress, Wf: fracture work.

The LPM compositions are given in Table 1. The fat ratio was chosen based on a previous study [30] to be consistent with the fat content of standard marketed cheeses. Pure water, milk fat, milk protein powder, and NaCl were vigorously mixed with a Blender® (Waring, Torrington, CT, USA) at medium speed (graduation 7) for 12 min. The mixture was cooled in a beaker, then placed in a thermostated bath at 32 °C. The pH was measured using a penetrometric electrode (Mettler-Toledo, Viroflay, France) and was adjusted at pH 6.5 using small aliquots of NaOH solution. After 30 min of resting, the aroma compound solution (0.5 mL) and rennet solution (4.8 mL) were added and mixed vigorously for 1 min. Prior to the coagulation, the mixture was immediately poured into a plastic bag and vacuum sealed, and then was completely immersed in a controlled-temperature bath at 32 °C for 3 h. The products were then stored at 4 °C until use.

- Instrumental texture analyses

LPM1 and LPM2 were analyzed by uniaxial compression at constant speed [31]. Cylinders of 3 cm length characterized by a ratio length/diameter of 1.3 were prepared for each LPM using a die-cutter of 2.3 cm internal diameter. Each sample was placed in a hermetically closed plastic cup at 19 °C for 15 min before experiments. The resistance strength, developed as a function of the deformation of the sample, was measured at 15 °C using a texture analyzer (TA-XT2, Stable Micro Systems Ltd., Champlan, France). The displacement speed of the superior plate (10 cm diameter) was 0.8 mm/s, and the samples were compressed until a deformation rate of 80% of their initial height.

2.2. *In Vivo Mastication*

Relevant physiological parameters (salivary flow rate, number of chew, shear angle, total duration of mastication, time between two chews, chew rate) were measured on three human subjects (volunteer staff of the laboratory; males; between 25 and 45 years old) without any pathology to obtain a range of representative and realistic values to use as inputs into the chewing simulator (Table 2). The procedure to analyze aroma release was the same as described in [32]. In particular, they were asked to chew the LPM cubes in their own way without any specific instruction. They were informed of the experiment objective and conditions. They gave their informed consent before they participated in the study. The study was conducted in accordance with the Declaration of Helsinki, and the protocol was approved by the relevant institutional and national regulations and legislation (Comité de Protection des Personnes Est-1, N° 2013/64–IDRCB 2013–A01084-41 on 21 November 2013).

Table 2. Physiological results obtained for each subject during chewing LPM1 and LPM2.

Subject	Product	Salivary Flow Rate (mL/s)	Chewing Duration (s)	Total Work of Muscle (mV s)	Number of Chewing Cycles	Mandible Force (daN)	Shearing Angle (°)
A	LPM1		13.5 ± 2.9	1.01 ± 0.31	25.0 ± 8.5	13.3 ± 1.1	2.6 ± 0.4
	LPM2	4.0 ± 0.4	15.1 ± 4.9	1.06 ± 0.13	21.3 ± 4.0	11.5 ± 0.5	2.7 ± 0.6
B	LPM1		23.1 ± 6.7	3.48 ± 0.25	33.7 ± 7.0	21.6 ± 1,5	5.0 ± 0.5
	LPM2	2.8 ± 0.2	24.9 ± 0.7	3.75 ± 0.23	38.0 ± 0	20.4 ± 0.5	4.7 ± 0.5
C	LPM1		28.0 ± 2.4	2.92 ± 0.19	39.0 ± 3.6	16.4 ± 0.4	3.1 ± 1.5
	LPM2	3.5 ± 0.2	23.7 ± 0.8	2.60 ± 0.16	33.7 ± 3.1	16.9 ± 0.8	3.0 ± 0.3

LPM1: lipoprotein matrix with 0.5 fat/milk protein ratio; LPM2: lipoprotein matrix with 1 fat/milk protein ratio; 3 replicates for each of the 3 subjects (A, B and C).

2.2.1. Salivary Flow Rate

The whole saliva flow rate was measured under real conditions of model cheese eating with LPM2, as it was reported in a previous study that salivary flow rate was not found to vary for the same subject when eating different model cheeses samples [5]. The total flow rate was determined by calculating the weight difference. To determine the salivation with both mechanical and chemical stimulations due to the presence of LPM in their mouths, the subjects were instructed to swallow just before introducing 5.0 ± 0.1 g of model cheese in their mouth and then to chew over a one minute period without swallowing. Next, the subjects spat out all the bolus and saliva in a tare flask, quickly cleansed their mouths with 10 g of mineral water, and spat in the same tare flask. All the measurements were performed in triplicate and at the same time of day for all subjects.

2.2.2. Electromyography Recording

The electromyography (EMG) recordings were conducted according to Mioche et al. [33]. The left and right superficial masseters and anterior temporalis muscles of each volunteer were located by palpation when the teeth were clenched. After careful cleaning of the overlying skin, two surface electrodes (Bionic France, Ternay, France) coated with a conductive paste were fixed 2 cm apart lengthwise along each muscle with an adhesive. An additional earth electrode was attached to the subjects' ear lobes. Each subject was instructed to naturally eat 5 g of model cheese. The experiments were performed in triplicate for each product.

After signal rectification, several variables were analyzed for the complete sequence of mastication, starting at the moment of food intake and ending at the last swallow—chewing time (total sequence duration before the last swallow), number of chews during the chewing time, chewing rate per minute, the mean voltage of each burst, the sum of the integrated areas of all individual bursts in the sequence (burst duration multiplied by its mean voltage expressed in mV·s), previously called muscle work, and the mean work (total work divided by the number of bursts).

2.2.3. Motion Capture

An infrared light-emitting diode (LED) was attached to the chin to follow the movements of the lower jaw, a second reference LED was attached to the brow, and a third one was attached to the laryngeal prominence to follow swallowing events. The LEDs were attached using a double-sided adhesive strap. The positions of the LEDs were recorded using an optical motion capture system (Northern Digital Optotrak, Radolfzell, Germany). The recording of jaw movement during chewing was superimposed to correctly align with the EMG recording to better repair chewing events. It was also used to calculate the average shearing angle of the lower jaw used as a parameter for the chewing simulator.

All the data gathered using these three methods were used as input parameters in the chewing simulator methods.

2.3. *In Vitro Mastication*

2.3.1. Chewing Simulator

A chewing simulator specially developed for in vitro flavor release study was used [24,28]. Particularly, it was fitted with a special valve for the online sampling of the gas phase. The chewing simulator was connected to an APCI-MS (mass spectrometer equipped with an atmospheric pressure ionization source) apparatus (Esquire-LC ion trap, Bruker Daltonique, Wissembourg, France). The operating procedure for mass spectrometry was the same as described in [32].

The upper part of the chewing simulator (Figure 1) was made of the upper jaw and conical palate containing the saliva inlet injector located approximately 10 mm from the upper teeth crown and central gas valve. The sampling valve was in close position when the distance between the tongue and palate was lower than 10 mm to avoid the valve and tubing from becoming contaminated and clogged. A bypass allowed the gas to flow towards the analyzer to maintain the baseline when the sampling valve was off. When the valve was open, volatiles in the headspace of the chewing simulator passed through the valve and were transferred to the mass spectrometer via the Venturi effect regulating the gas flow rate. On the opposite side of the valve, a vent allowed external air to enter the cell. All the surfaces of the chewing simulator in contact with the food were made of polyetheretherketone (PEEK). The transfer line was a deactivated fused silica capillary tube [24,28].

The bite force (BF; between 20 and 35 daN), salivary flow rate (between 1 and 4 mL/min), and shearing angle (SA; 3 and 5°, respectively, corresponding to a 1/8 and 1/5 tooth shift, respectively) were the variable parameters. Three replicates were performed for each measurement condition and for each of the two LPMs (Table 1). Artificial saliva containing minerals and mucin was used [34]. A syringe pump controlled by the chewing simulator software ensured the constant artificial salivary flow rate.

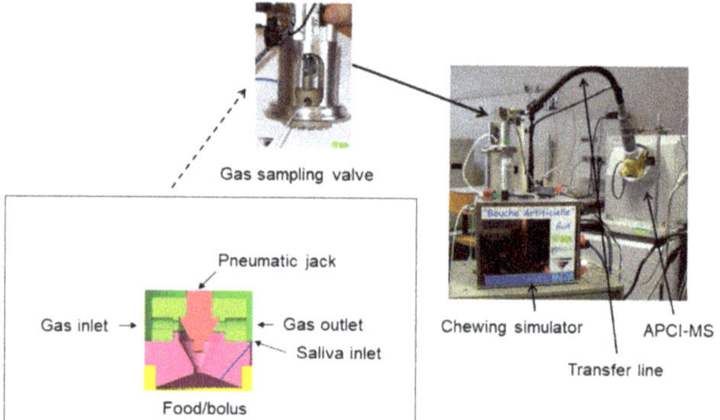

Figure 1. Chewing simulator connected to the atmospheric pressure chemical ionization mass spectrometer (APCI-MS) for online and real time analyses of aroma compound release under in vitro mastication conditions with details of the gas sampling valve.

2.3.2. Volatile Compound Release Measurements

The nosespace sampled online from a subject or the chewing simulator was drawn at 30 or 80 mL/min and flowed through a heated capillary transfer line (150 °C, 0.53 mm i.d.) into an ion-trap mass spectrometer Esquire-LC (Brucker Daltronique, Wissembourg, France) equipped with a modified APCI source [35]. Aroma compounds were ionized by a 5-kV positive ion corona pin discharge. After checking the main ions for each volatile compound using single aroma compound solutions as references, the data were collected for detectable ions corresponding to protonated aroma compound molecules incorporated in the model cheeses (Table 3). The intensity of the signal was expressed in arbitrary units. In all cases, the room air and exhaled air by the subject or by the chewing simulator—before introducing the food sample in the mouth—was recorded and then was subtracted from the records obtained when chewing foods. Aroma release, jaw muscle contraction, and motion were recorded at the same time for the three subjects. Two parameters were extracted from each release curve—the release rate (RR) taken at the beginning of the chewing process, and magnitude, defined as the difference between the level of release before and after the chewing process (Cmax) (Figure 2).

Table 3. In vivo aroma compound release. Fisher statistics from Analyses of Variance (ANOVA) of the temporal aroma compound release parameters. Only significant results ($p < 0.05$) are reported. Cmax: maximum released concentration; RR: release rate; LPM: lipoprotein matrix.

Ions (m/z)	Release Parameters	F	p	Subject Effect
89	Cmax	5.66	0.018	B > (C = A)
	RR	7.44	0.007	B > (C = A)
115	Cmax	7.71	0.007	(B = C) > A
	RR	9.36	<0.001	B > (C = A)
117	Cmax	5.47	0.021	B ≥ C ≥ A (B > A)
	RR	5.44	0.021	B ≥ A ≥ C (B > C)
129	Cmax	5.98	0.015	C ≥ B ≥ A (C > A)
	RR	7.98	0.006	(B = A) > C
143	Cmax	5.88	0.016	(B = C)> A
	RR	7.34	0.008	B ≥ A ≥ C (B > C)
	RR	9.31	0.003	Subject*LPM: B-LPM2 > B-LMP1
145	RR	4.58	0.033	B ≥ A ≥ C (B > C)

m/z: mass/charge ratio; m/z 89 (Butanoic acid); m/z 115 (2-Heptanone); m/z 117 (Ethyl butanoate); m/z 129 (3-Octanone); m/z 143 (2-Nonanone); m/z 145 (Ethyl hexanoate). F: Fisher value calculated by ANOVA; * indicates interaction between parameters.

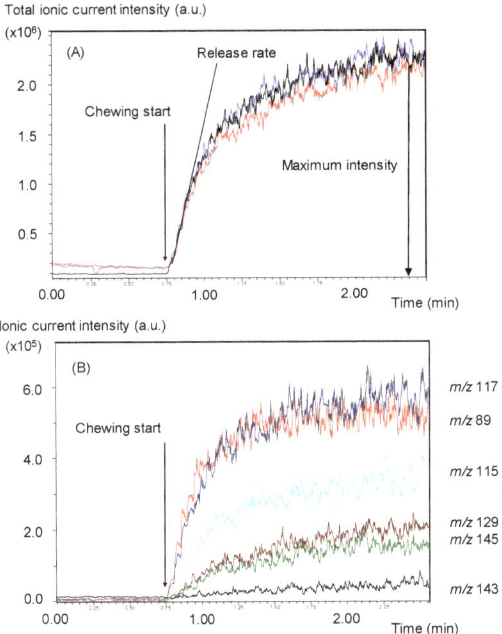

Figure 2. Example of in vitro temporal aroma compound release curves obtained by online coupling of the chewing simulator with the APCI-MS apparatus (3 replicates). (**A**) The total ionic current presented in ordinate is representative of the overall volatile compounds concentration in the gas phase according to the time. (**B**) Ionic current intensity for each detected ion in function of the chewing time. Conditions: initial volume of saliva: 1 mL; salivary flow rate: 1 mL/min; compression force: 25 DaN; shearing angle: 3°. Volatile compounds in lipoprotein matrices (μL/kg): butanoic acid (10); 2,3-butanedione (3); 2-heptanone (5); 2-nonanone (5); 3-octanone (5); ethyl hexanoate (5); ethyl butanoate (4); dimethyldisulphide (10). Ions: *m/z* 117 (ethylbutanoate); *m/z* 89 (butanoic acid); *m/z* 115 (2-heptanone); *m/z* 129 (3-octanone); *m/z* 145 (ethylhexanoate); *m/z* 143 (2-nonanone).

The hydrophobicity of the aroma compounds (Log P) was calculated with BIOVIA discovery studio software (version 2017, BIOVIA Corporate Europe, Cambridge, UK) according to Ghose et al. [35].

2.4. Statistical Analyses

The data were processed using Stat SAS system release 6.12 (SAS Institute Inc, Cary, NC, USA). Analyses of Variance (ANOVA) were performed at the level $\alpha = 0.05$ to study the variation in oral parameters (model including product, subject, and product * subject effects) and aroma compound release parameters (model including product, oral parameter, product * oral parameter). The means were compared using the Newman-Keuls multiple comparison tests.

3. Results

The two lipoprotein matrices used in this study were different only in their fat/protein ratio (0.5 and 1). These food matrices were comprised of milk fat and proteins that were coagulated using rennet. Their composition and rheological characteristics are given in Table 1. Among the four rheological parameters obtained by uniaxial compression, significant differences were observed for the modulus of deformability (MD). Thus, the lipoprotein matrix LPM2 with higher dry matter and fat content showed a higher MD value, indicating a harder structure than LPM1. The LPMs were aromatized with a very simple volatile compound composition to mimic a typical cheese flavor (Table 1). However, two of them,

2,3-butanedione and dimethyldisulphide, were not detectable by APCI-MS because of the low sensitivity for their detection. Thus, the presented data concern only the five detectable ions corresponding to butanoic acid, 2-heptanone, 2-nonanone, 3-octanone, ethyl hexanoate, and ethyl butanoate (Table 1).

3.1. In Vivo Aroma Compound Release

The oral physiological characteristics of the three subjects (named A, B, and C) were evaluated mainly to limit the decoupled oral functions of the chewing simulator to simulate mastication and salivation with realistic values. The obtained results with the three subjects were in agreement with previous works [5,20,21], thus we considered that the obtained results could be used as relevant information for the in vitro experiments. These in vivo results are briefly presented below.

The oral physiological parameters recorded on the three subjects are reported in Table 2. No influence of LPM or the interaction subject LPM was observed on chewing parameters. The influence of subjects was observed on most of the measured oral parameters. The salivary flow rate was evaluated for each panelist using a lipoprotein matrix (LPM2) to obtain representative values with the food product, considering mechanical and chemical stimulations specifically provided by the studied products. The salivary flow rates of A and B were significantly higher than that of C. By contrast, the shearing angles of A and B were significantly lower than that of C. The number of chewing cycles and chewing duration of B and C were similar but significantly higher than those of A. The maximal amplitude of chewing cycle and total muscle work were significantly different for each of the subjects in the decreasing order: B > C > A.

Concerning aroma compound release, no influence of the LPM on either the RR value or on Cmax—regardless of the ion considered—was observed. The overall effect of the subjects on RR was significant ($F = 11.94$, $p = 0.00002$) with the decreasing order: B > C > A.

The effects of the subject factors on Cmax and RR and interaction with LPM are reported in Table 3, confirming the large inter-individual variability. Correlations between the aroma release and chewing parameters were found. Cmax and the number of chewing cycles were well correlated regardless of the ion considered. For each ion, the significant correlations were as follows: butanoic acid (m/z 89; $r = 0.48$, $p = 0.041$), 2-heptanone (m/z 115; $r = 0.68$, $p = 0.002$), ethyl butyrate (m/z 117; $r = 0.57$, $p = 0.012$), 3-octanone (m/z 129; $r = 0.55$, $p = 0.018$), 2-nonanone (m/z 143; $r = 0.58$, $p = 0.011$), and ethyl hexanoate (m/z 145; $r = 0.47$, $p = 0.044$). Correlations between Cmax and total muscle work for ions 89 ($r = 0.67$, $p = 0.002$), 115 ($r = 0.52$, $p = 0.027$), 117 ($r = 0.67$, $p = 0.002$), and 143 ($r = 0.578$, $p = 0.012$) were also reported. Moreover, for ion 89, correlations between RR and total muscle work ($r = 0.54$, $p = 0.021$) and between RR and the salivary flow rate ($r = -0.72$, $p = 0.026$), were observed. However, because of the low number of observations, these in vivo data could not be further interpreted and should be further validated with a larger number of panelists. They were mainly used as a basis for our choice of in vitro chewing parameters.

3.2. In Vitro Chewing Process

The good superposition of the total ionic current obtained for three replicates made under the same conditions shows the good reproducibility of the in vitro aroma compound release measurements made by direct coupling between the chewing simulator and APCI-MS. For example, three replicates obtained with LPM1 are shown in Figure 2. The reproducibility was generally better than what can be observed in in vivo studies that are sensitive to inter- and intra-individual variability and reproducibility [36,37]. The pattern of the temporal curve obtained while using the device was different from the typical symmetrical curve, with a pre-swallowing and a post-swallowing phase that could usually be observed for in vivo temporal compound release kinetics [38]. On the in vitro curves, the absence of the post-swallowing phase was due to the absence of a swallowing function of the chewing simulator. Thus, only the release rate (RR) and maximum concentration of released compounds (Cmax) could be extracted from the pre-swallowing phase of the curve.

The release kinetics for each detectable ion and their analyses were followed according to their volatility and affinity for the different phases after extraction of the initial slope, indicating the release rate and maximum intensity from these curves. If each ion was considered independently, it was clear that differences in aroma compound release were observed according to the physicochemical characteristics of the volatile compound and the oral parameters. For each ion, all the obtained results are reported in Table 4. The maximum concentration values of less hydrophobic compounds (butanoic acid, 2-heptanone, and ethyl butyrate) were mainly affected by bite force and shear angle. A higher bite force and shear angle resulted in a higher aroma release. The aroma release parameters of more hydrophobic compounds (3-octanone, 2-nonanone, and ethyl hexanoate) were, in general, affected by the salivary flow rate, but the effect depended on the type of lipoprotein matrix and shear angle, as observed by the significant interaction effects.

4. Discussion

4.1. Texture of the Lipoprotein Matrices

The two LPMs mainly varied in their fat/milk protein ratio by a factor of two. This parameter is an important factor in aroma compound release and texture variability [4]. The rheological properties of the two LPMs presented in Table 1 only differed in their modulus of deformability, indicating a lower elasticity of LPM2 than that of LPM1. Although the LPMs required sufficient chewing activity to be broken in the mouth, their difference in elasticity did not affect the oral parameters of the panelists (Table 2). It was noticed that, with a larger variation in texture, differences in the oral parameters were reported [4,39,40], and during all the in vivo process, effects of bolus hardness on masticatory kinetics were observed [2,41,42]. Thus, the absence of a difference in the oral parameters between the two LPMs for each panelist could be explained by the difference in texture between the two LPMs being insufficient to trigger to a change in the mastication pattern of the subjects.

4.2. Effect of Bite and Shear Forces on In Vitro Aroma Compound Release

Concerning in vitro aroma compound release, mechanical mastication increased the release of the aroma compounds, as already mentioned by several works performed with in vitro systems [22,43,44].

In our study (Table 4), the less hydrophobic volatile compounds—butanoic acid, 2-heptanone, and ethyl butanoate—were more released when the bite force and shearing angle were higher (Table 4). This showed the importance of these two mechanical parameters for the release of these compounds. The progressive increase in the exchange area of the particles due to the breakdown of the matrix could easily explain this observation. A high bite force led to better food breakdown in the mouth. However, no effect on the release rate was found in either BF or SA, suggesting that these two parameters did not influence the kinetics of release of the less hydrophobic volatile compounds but only the overall quantity of the released components.

For the more hydrophobic compounds, the effects of BF and SA were less clear and seemed more specific to the nature of the volatile compound. Concerning the maximum released concentration, a significant effect of BF (but not of SA) was observed for 3-octanone, while a significant effect of SA (but not of BF) was observed for 2-nonanone. For ethyl hexanoate, only a significantly higher release rate was observed with an increase in the shearing angle. These volatile compound behavior differences were rather difficult to interpret and need further specific investigations to be more clearly understood. The data reported in Table 4 clearly show the impact of the shearing angle in the aroma compound release intensity. Generally, the overall mechanical effect of the jaw on food breakdown and aroma compound release is the result of shear and compression forces. In this study, shear force, related to the horizontal movement, could be independent of the bite force related to the vertical movement of the chewing functions reproduced in vitro, showing their own impact on aroma compound release. The closing angle and width of the chewing loop were found to be significantly correlated with the toughness and modulus of elasticity, which were two mechanical properties of the food, whereas the vertical movement was not correlated [45]. In addition, it was reported that, in vivo, the lateral chewing motion was dependent on the consistency of the food, although differences between subjects were observed due to the retro control by mechanoreceptors [46]. A greater consistency of foods resulted in a larger lateral chewing motion of the lower jaw. These observations were in favor of a high involvement of the shearing forces in the food breakdown of hard foods and bolus formation, leading to higher spreadability of the bolus and potentially higher emulsification of the bolus in saliva, thus increasing the exchange surface with air. This could influence aroma compound release differently according to the nature of the volatile compounds.

Table 4. In vitro aroma compound release from lipoprotein matrices. Fisher statistics from Analyses of Variance (ANOVA) of the temporal aroma compound release parameters. Only significant results ($p < 0.05$) are reported. Cmax: maximum released concentration; RR: release rate; LPM: lipoprotein matrix; BF: bite force; SA: shearing angle; SF: salivary flow rate.

Compound (Log P)		**Butanoic Acid (0.918)**		**Ethyl Butanoate (1.492)**		**2-Heptanone (1.931)**		**Ethyl Hexanoate (2.405)**		**3-Octanone (2.598)**		**2-Nonanone (2.843)**	
m/z		89		115		117		145		129		143	
		Cmax	RR	Cmax	RR	Cmax	RR	Cmax	RR	Cmax	RR	Cmax	RR
BF	F	15.77		7.09		5.02						26.56	
	p-value	0.0005		0.0139		0.0356						<0.0001	
SA	F	23.10		12.51		8.92			9.17	5.22			
	p-value	<0.0001		0.0017		0.0064			0.0058	0.0315			
SF	F							13.89	7.24	13.30		6.30	
	p-value							<0.0001	0.0035	0.0001		0.0063	
LPM*SA	F						6.04						
	p-value						0.0216						
LPM*SF	F				6.11			7.19		6.20			6.95
	p-value				0.0213			0.031		0.0201			0.0145
SF*SA	F							5.59	5.55	5.62			5.97
	p-value							0.0101	0.0105	0.0099			0.0078

m/z: mass/charge ratio; Log P: hydrophobicity of the aroma compounds (calculated with BIOVIA™ discovery studio software (version 2017, BIOVIA™ Corporate Europe, Cambridge, UK) according to Ghose et al. [47]. F: Fisher value calculated by ANOVA; * indicates interaction between parameters.

4.3. *Effect of Salivary Flow Rate on In Vitro Aroma Compound Release*

A significant effect of the salivary flow rate on the aroma compound release was observed only for the most hydrophobic compounds—ethyl hexanoate, 3-octanone, and 2-nonanone. For these compounds, an increase in the salivary flow rate led to a decrease in the maximum concentration of released compounds, and in addition, a decrease in the release rate for ethyl hexanoate, which was the least hydrophobic of these three molecules. We have to consider here that the food matrix is a complex medium. Consequently, the behavior of volatile compounds and their partition between the different phases cannot be explained only by their hydrophobicity, as is the case for simple models. Despite the hydrophobic character of these compounds, the progressive increase of artificial saliva volume should be in favor of their displacement to the water phase, leading to a higher release in the air phase because of their hydrophobicity [48]. Although the retention/release phenomenon is not understood completely in complex food matrices [49], a possible explanation is the effect of dilution by an increase of saliva volume, increasing the capacity of the water phase to retain more hydrophobic volatile compounds, leading to a partition less in favor of the gas phase. This effect does not affect the less hydrophobic molecules that are more soluble in water. It cannot be excluded that this effect could also be due to interactions with salivary proteins that are progressively introduced with saliva during the chewing process, but this is not generally described as the major phenomenon that explains volatile compound retention [12]. Previous in vitro studies have shown that increasing salivary flow rate decreased the aroma released from bell peppers [44], and the addition of saliva and water decreased the release of some aroma compounds from coffee brews [50].

However, the impact of the salivary flow rate on aroma compound release is not so clear and varies according to the studies and conditions. In an in vivo study, Pionnier et al. found no influence of the salivary flow rate on the release of aroma compounds of various polarities during the chewing of a model cheese [38]. In another in vivo study, it was reported that the salivary flow rate did not vary among cheese samples, but, at the end of mastication, the amount of saliva in boluses differed depending on the hardness and fat content of cheeses [5]. The authors suggested that the combined effects of both the chewing process and salivary flow rates were responsible for differences in aroma compound release between subjects. However, it was not possible to independently decouple the effects of each phenomenon.

4.4. Effect of Oral Functions and Food Composition Interactions on In Vitro Aroma Compound Release

Different interactions between oral parameters and composition factors were observed, thus modulating the main effects described above.

Although fewer effects were observed concerning the release rate, a significant interaction showed that, only for 2-heptanone, the release rate was significantly higher when the shearing angle and fat content were the highest, also corresponding to the hardest lipoprotein matrix, which contained less water and fewer proteins. This effect seemed rather complex and specific to this compound.

For the more hydrophobic volatile compounds, salivary flow rate effect showed significant interaction with the LPM and shear angle factors, indicating that the effect salivary flow rate had on aroma compound release depended on the food matrix and the extent of chewing. The effect of the salivary flow rate showed interaction with other factors. Significant interaction with the lipoprotein matrix was observed for the maximum concentration of the more hydrophobic aroma compound released—apart from 2-nonanone, for which the interaction was observed for RR. The maximum concentrations of the released 3-octanone and ethyl hexanoate were significantly higher at the high salivary flow rate when the fat content was highest, while it was the opposite for ethyl butanoate, which was the less hydrophobic volatile compound for which this interaction was significant. 2-nonanone was released significantly more slowly when the food matrix contained more proteins and less fat, and the salivary flow rate was low. Slight differences occurred between the effects of interactions between the volatile compounds. However, the resulting observed effect was that the combination of the low salivary flow rate and low fat and high protein content of the food matrix was in favor of a lower release of hydrophobic volatile compounds, while the combination of a high salivary flow rate and high fat and low protein food matrix was in favor of a higher release. These observations were consistent with previous observations [4,51]. However, this interaction was not easy to interpret because several composition parameters varied. This needs further in vitro experiments to be unraveled.

In in vivo studies, it is often questionable whether such effects of texture and composition of the food matrices on aroma compound release are due to physicochemical interactions between the food components and volatile compounds or due to a change in the chewing behavior because of the retro-control of the chewing function by texture perception [39,40]. Here, the control and decoupling of the chewing parameters using the device allowed us to discard the second assessment.

Significant interactions between the salivary flow rate and shearing angle were also observed for the release of the more hydrophobic volatile compounds. For 3-octanone and ethyl hexanoate, the maximum concentration of released compounds was significantly higher for the higher salivary flow rate and shearing angle than for the other cases due to a more efficient extraction with an important mixing effect in a larger quantity of solvent. A lower release rate of 2-nonanone and ethyl hexanoate was observed when low shearing and low salivary flow rate were combined. In this case, a lower mixing combined with a low salivary flow rate led to fewer exchanges between the gas phase and liquid phase and thus to a slower release.

5. Conclusions

The chewing simulator can precisely reproduce the compression and shear forces of human jaws, causing temporal food breakdown and compound release. In this in vitro study, each oral parameter and the interactions between them could differently influence temporal aroma compound release according to the nature of the aroma compounds. Moreover, we were able to control each parameter separately so that the effects could be measured one at a time in isolation, particularly for the bite and shear forces on aroma compound release.

In most in vivo studies dealing with relationships between oral processing and temporal aroma compound release, it is difficult to attribute the relative importance of each single oral parameter to the individual aroma compound release pattern. Moreover, individuals adapt their individual eating pattern to the texture and consistency of the food matrix, leading to changes in aroma compound release [5,52]. Thus, generally, the part of the in-mouth release of stimuli due to food characteristics and the adaptation of food oral processing according to texture perception are difficult to differentiate. The possibility of controlling and decoupling oral functions with our chewing simulator device could allow overcoming limitations of the in vivo aroma compound release studies.

Using this device, we were able to explain the temporal release of the studied volatile compounds by the effect of the components of the surrounding media and physical constraints due to mastication on aroma compound release. In particular, we reported effects of salivary flow rate varying according to the hydrophobicity of the volatile compounds and a clear effect of bite force and shear angle on the less hydrophobic volatile compounds, leading to a higher release. The disconnection between the bite force and the shearing angle allowed us to highlight the importance of the shear angle on flavor release during chewing. However, for the release of the more hydrophobic volatile compounds, the effects of the bite force and shear angle seemed more typical of the nature of the compounds and thus were more difficult to interpret. Significant interactions were also observed between shear angle, salivary flow rate, and lipoprotein matrix composition, mainly for the release of the more hydrophobic volatile compounds, which needs further investigation to be clarified.

Thus, this tool allows identification of the main physical or physiological phenomena explaining the active release, as well as validation of the proposed assumptions to model in silico aroma compound release. Additional functionalities that have been shown to be important for aroma compound release are lacking. Their integration in this device and the improvement of the system are under investigation.

Author Contributions: Data curation, A.T., C.Y. and E.S.; Investigation, A.T. and C.Y.; Methodology, E.S. and P.M.; Supervision, C.S.; Writing-original draft, C.S.; Writing-review and editing, A.T., C.Y., E.S., P.M. and C.S.

Funding: This research was funded by INRA (Internal); the Conseil Régional de Bourgogne (France) and European Regional Development Fund (ERDF) funded the equipment used in this project (development of chewing simulator, mass spectrometer).

Acknowledgments: We sincerely thank the Plateform of Technology (Plateform3D, SAYENS, Le Creusot, France) for developing the device and their technical help. We acknowledge Jacques Maratray (CSGA, INRA, Dijon, France) for developing the software to monitor the chewing simulator and Christine Achilléos (URTAL, INRA, Poligny, France) for the rheological measurements.

Conflicts of Interest: The authors declare no conflict of interest. The funders had no role in the design of the study; in the collection, analyses, or interpretation of data; in the writing of the manuscript, or in the decision to publish the results.

References

1. Salles, C.; Chagnon, M.C.; Feron, G.; Guichard, E.; Laboure, H.; Morzel, M.; Semon, E.; Tarrega, A.; Yven, C. In-mouth mechanisms leading to flavor release and perception. *Crit. Rev. Food Sci. Nutr.* **2011**, *51*, 67–90. [CrossRef] [PubMed]
2. Chen, J. Food oral processing—A review. *Food Hydrocolloid.* **2009**, *23*, 1–25. [CrossRef]
3. Deleris, I.; Saint-Eve, A.; Semon, E.; Guillemin, H.; Guichard, E.; Souchon, I.; Le Quere, J.-L. Comparison of direct mass spectrometry methods for the on-line analysis of volatile compounds in foods. *J. Mass Spectrom.* **2013**, *48*, 594–607. [CrossRef] [PubMed]
4. Boisard, L.; Tournier, C.; Sémon, E.; Noirot, E.; Guichard, E.; Salles, C. Salt and fat contents influence the microstructure of model cheeses, chewing/swallowing and in vivo aroma release. *Flav. Fragr. J.* **2014**, *29*, 95–106. [CrossRef]
5. Tarrega, A.; Yven, C.; Sémon, E.; Salles, C. In-mouth aroma compound release during cheese consumption: Relationship with food bolus formation. *Int. Dairy J.* **2011**, *21*, 358–364. [CrossRef]

6. Frank, D.; Appelqvist, I.; Piyasiri, U.; Delahunty, C. In vitro measurement of volatile release in model lipid emulsions using proton transfer reaction mass spectrometry. *J. Agric. Food Chem.* **2012**, *60*, 2264–2273. [CrossRef] [PubMed]
7. Guichard, E. Interactions between flavor compounds and food ingredients and their influence on flavor perception. *Food Rev. Int.* **2002**, *18*, 49–70. [CrossRef]
8. Paravisini, L.; Guichard, E. Interactions between aroma compounds and food matrix. In *Flavour: From Food to Perception*; Guichard, E., Salles, C., Morzel, M., Le Bon, A.-M., Eds.; John Wiley & Sons Inc.: Chichester, UK, 2017; pp. 208–234.
9. Rekker, R.F. *The Hydrophobic Fragmental Constant. Its Derivation and Application*; Nauta, W., Rekker, R.F., Eds.; Elsevier Scientific Publishing Co.: Amsterdam, The Netherlands, 1977; pp. 1–389.
10. Frank, D.; Eyres, G.T.; Piyasiri, U.; Cochet-Broch, M.; Delahunty, C.M.; Lundin, L.; Appelqvist, I.M. Effects of agar gel strength and fat on oral breakdown, volatile release, and sensory perception using in vivo and in vitro systems. *J. Agric. Food Chem.* **2015**, *63*, 9093–9102. [CrossRef] [PubMed]
11. Frank, D.C.; Eyres, G.T.; Piyasiri, U.; Delahunty, C.M. Effect of food matrix structure and composition on aroma release during oral processing using in vivo monitoring. *Flav. Fragr. J.* **2012**, *27*, 433–444. [CrossRef]
12. Ployon, S.; Morzel, M.; Canon, F. The role of saliva in aroma release and perception. *Food Chem.* **2017**, *226*, 212–220. [CrossRef]
13. Hirtz, C.; Chevalier, F.; Centeno, D.; Rofidal, V.; Egea, J.C.; Rossignol, M.; Sommerer, N.; de Periee, D.D. MS characterization of multiple forms of alpha-amylase in human saliva. *Proteomics* **2005**, *5*, 4597–4607. [CrossRef] [PubMed]
14. Perry, G.H.; Dominy, N.J.; Claw, K.G.; Lee, A.S.; Fiegler, H.; Redon, R.; Werner, J.; Villanea, F.A.; Mountain, J.L.; Misra, R.; et al. Diet and the evolution of human amylase gene copy number variation. *Nat. Genet.* **2007**, *39*, 1256–1260. [CrossRef] [PubMed]
15. Denny, P.C.; Denny, P.A.; Klauser, D.K.; Hong, S.H.; Navazesh, M.; Tabak, L.A. Age-related-changes in mucins from human whole saliva. *J. Dent. Res.* **1991**, *70*, 1320–1327. [CrossRef]
16. Feron, G. Unstimulated saliva: Background noise in taste molecules. *J. Texture Stud.* **2018**. [CrossRef] [PubMed]
17. Mejean, C.; Morzel, M.; Neyraud, E.; Issanchou, S.; Martin, C.; Bozonnet, S.; Urbano, C.; Schlich, P.; Hercberg, S.; Peneau, S.; et al. Salivary Composition Is Associated with Liking and Usual Nutrient Intake. *PloS ONE* **2015**, *10*. [CrossRef]
18. Neyraud, E.; Palicki, O.; Schwartz, C.; Nicklaus, S.; Feron, G. Variability of human saliva composition: Possible relationships with fat perception and liking. *Arch. Oral Biol.* **2012**, *57*, 556–566. [CrossRef]
19. Guichard, E.; Salles, C.; Morzel, M.; Le Bon, A.-M. *Flavour: From Food to Perception*; John Wiley & Sons Inc.: Chichester, UK; Hoboken, NJ, USA, 2017; pp. 1–400.
20. Gierczynski, I.; Laboure, H.; Guichard, E. In vivo aroma release of milk gels of different hardnesses: Inter-individual differences and their consequences on aroma perception. *J. Agric. Food Chem.* **2008**, *56*, 1697–1703. [CrossRef]
21. Laboure, H.; Repoux, M.; Courcoux, P.; Feron, G.; Guichard, E. Inter-individual retronasal aroma release variability during cheese consumption: Role of food oral processing. *Food Res. Int.* **2014**, *64*, 692–700. [CrossRef]
22. Poinot, P.; Arvisenet, G.; Grua-Priol, J.; Fillonneau, C.; Prost, C. Use of an artificial mouth to study bread aroma. *Food Res. Int.* **2009**, *42*, 717–726. [CrossRef]
23. Salles, C.; Benjamin, O. Models of the oral cavity for the investigation of olfaction. In *Handbook of Odour*; Buettner, A., Ed.; Springer: Berlin, Germany, 2017; pp. 303–318.
24. Salles, C.; Tarrega, A.; Mielle, P.; Maratray, J.; Gorria, P.; Liaboeuf, J.; Liodenot, J.J. Development of a chewing simulator for food breakdown and the analysis of in vitro flavor compound release in a mouth environment. *J. Food Eng.* **2007**, *82*, 189–198. [CrossRef]
25. Piggott, J.R.; Schaschke, C.J. Release cells, breath analysis and in-mouth analysis in favour research. *Biomol. Eng.* **2001**, *17*, 129–136. [CrossRef]
26. Krause, A.J.; Henson, L.S.; Reineccius, G.A. Use of a chewing device to perform a mass balance on chewing gum components. *Flav. Fragr. J.* **2011**, *26*, 47–54. [CrossRef]

27. Benjamin, O.; Silcock, P.; Kieser, J.A.; Waddell, J.N.; Swain, M.V.; Everett, D.W. Development of a model mouth containing an artificial tongue to measure the release of volatile compounds. *Innov. Food Sci. Emerg. Technol.* **2012**, *15*, 96–103. [CrossRef]
28. Mielle, P.; Tarrega, A.; Sémon, E.; Maratray, J.; Gorria, P.; Liodenot, J.J.; Liaboeuf, J.; Andrejewski, J.L.; Salles, C. From human to artificial mouth, from basics to results. *Sens. Actuator B Chem.* **2010**, *146*, 440–445. [CrossRef]
29. Floury, J.; Camier, B.; Rousseau, F.; Lopez, C.; Tissier, J.-P.; Famelart, M.-H. Reducing salt level in food: Part 1. Factors affecting the manufacture of model cheese systems and their structure-texture relationships. *Lwt-Food Sci. Technol.* **2009**, *42*, 1611–1620. [CrossRef]
30. Lawrence, G.; Buchin, S.; Achilleos, C.; Berodier, F.; Septier, C.; Courcoux, P.; Salles, C. In vivo sodium release and saltiness perception in solid lipoproteic matrices. 1. Effect of composition and texture. *J. Agric. Food Chem.* **2012**, *60*, 5287–5298. [CrossRef]
31. Noel, Y.; Zannoni, M.; Hunter, E.A. Texture of Parmigiano Reggiano cheese: Statistical relationships between rheological and sensory variates. *Lait* **1996**, *76*, 243–254. [CrossRef]
32. Tarrega, A.; Yven, C.; Semon, E.; Salles, C. Aroma release and chewing activity during different model cheeses. *Int. Dairy J.* **2008**, *18*, 849–857. [CrossRef]
33. Mioche, L.; Bourdiol, P.; Martin, J.F.; Noël, Y. Variations in human masseter and temporalis muscle activity related to food texture during free and side-imposed mastication. *Arch. Oral Biol.* **1999**, *44*, 1005–1012. [CrossRef]
34. Odake, S.; Roozen, J.P.; Burger, J.J. Flavor release of diacetyl and 2-heptanone from cream style dressings in three mouth model systems. *Biosci. Biotechnol. Biochem.* **2000**, *64*, 2523–2529. [CrossRef]
35. Sémon, E.; Gierczynski, I.; Langlois, D.; Le Quéré, J.-L. Analysis of aroma compounds by atmospheric pressure chemical ionisation—Ion trap mass spectrometry. Construction and validation of an interface for in vivo analysis of human breath volatile content. In Proceedings of the 16th International Mass Spectrometry Conference, Edinburgh, Scotland, UK, 31 August–5 September 2003; CD-ROM Supplement, abstract 324.
36. Repoux, M.; Semon, E.; Feron, G.; Guichard, E.; Laboure, H. Inter-individual variability in aroma release during sweet mint consumption. *Flav. Fragr. J.* **2012**, *27*, 40–46. [CrossRef]
37. Ruijschop, R.M.A.J.; Burgering, M.J.M.; Jacobs, M.A.; Boelrijk, A.E.M. Retro-Nasal Aroma Release Depends on Both Subject and Product Differences: A Link to Food Intake Regulation? *Chem. Senses* **2009**, *34*, 395–403. [CrossRef]
38. Pionnier, E.; Chabanet, C.; Mioche, L.; Le Quere, J.L.; Salles, C. 1. In vivo aroma release during eating of a model cheese: Relationships with oral parameters. *J. Agric. Food Chem.* **2004**, *52*, 557–564. [CrossRef]
39. Dan, H.; Kohyama, K. Interactive relationship between the mechanical properties of food and the human response during the first bite. *Arch. Oral Biol.* **2007**, *52*, 455–464. [CrossRef]
40. van der Bilt, A. Assessment of mastication with implications for oral rehabilitation: A review. *J. Oral Rehabil.* **2011**, *38*, 754–780. [CrossRef]
41. Anderson, K.; Throckmorton, G.S.; Buschang, P.H.; Hayasaki, H. The effects of bolus hardness on masticatory kinematics. *J. Oral Rehabil.* **2002**, *29*, 689–696. [CrossRef]
42. Chen, J. Food oral processing: Mechanisms and implications of food oral destruction. *Trends Food Sci. Technol.* **2015**, *45*, 222–228. [CrossRef]
43. van Ruth, S.M.; Buhr, K. Influence of mastication rate on dynamic flavour release analysed by combined model mouth/proton transfer reaction-mass spectrometry. *Int. J. Mass Spectrom.* **2004**, *239*, 187–192. [CrossRef]
44. Van Ruth, S.M.; Roozen, J.P. Influence of mastication and saliva on aroma release in a model mouth system. *Food Chem.* **2000**, *71*, 339–345. [CrossRef]
45. Agrawal, K.R.; Lucas, P.W.; Bruce, I.C. The effects of food fragmentation index on mandibular closing angle in human mastication. *Arch. Oral Biol.* **2000**, *45*, 577–584. [CrossRef]
46. Filipic, S.; Keros, J. Dynamic influence of food consistency on the masticatory motion. *J. Oral Rehabil.* **2002**, *29*, 492–496. [CrossRef] [PubMed]
47. Ghose, A.K.; Viswanadhan, V.N.; Wendoloski, J.J. Prediction of hydrophobic (lipophilic) properties of small organic molecules using fragmental methods: An analysis of ALOGP and CLOGP methods. *J. Phys. Chem. A* **1998**, *102*, 3762–3772. [CrossRef]

48. Tromelin, A.; Andriot, I.; Kopjar, M.; Guichard, E. Thermodynamic and Structure Property Study of Liquid-Vapor Equilibrium for Aroma Compounds. *J. Agric. Food Chem.* **2010**, *58*, 4372–4387. [CrossRef] [PubMed]
49. Ayed, C.; Lubbers, S.; Andriot, I.; Merabtine, Y.; Guichard, E.; Tromelin, A. Impact of structural features of odorant molecules on their retention/release behaviours in dairy and pectin gels. *Food Res. Int.* **2014**, *62*, 846–859. [CrossRef]
50. Genovese, A.; Caporaso, N.; Civitella, A.; Sacchi, R. Effect of human saliva and sip volume of coffee brews on the release of key volatile compounds by a retronasal aroma simulator. *Food Res. Int.* **2014**, *61*, 100–111. [CrossRef]
51. Boisard, L.; Andriot, I.; Martin, C.; Septier, C.; Boissard, V.; Salles, C.; Guichard, E. The salt and lipid composition of model cheeses modifies in-mouth flavour release and perception related to the free sodium ion content. *Food Chem.* **2014**, *145*, 437–444. [CrossRef] [PubMed]
52. Blissett, A.; Hort, J.; Taylor, A.J. Influence of chewing and swallowing behavior on volatile release in two confectionery systems. *J. Texture Stud.* **2006**, *37*, 476–496. [CrossRef]

Article

Potential Impact of Oat Ingredient Type on Oral Fragmentation of Biscuits and Oro-Digestibility of Starch—An In Vitro Approach

Amparo Gamero [1], Quoc Cuong Nguyen [2], Paula Varela [2], Susana Fiszman [1,*], Amparo Tarrega [1] and Arantxa Rizo [1]

[1] Instituto de Agroquímica y Tecnología de Alimentos (IATA-CSIC), 46980 Valencia, Spain; agamero@iata.csic.es (A.G.); atarrega@iata.csic.es (A.T.); arantxa.rizo@iata.csic.es (A.R.)
[2] Nofima AS. NO-1433 Ås Norway; quoc.cuong.nguyen@nofima.no (Q.C.N.); paula.varela.tomasco@nofima.no (P.V.)
* Correspondence: sfiszman@iata.csic.es

Received: 6 April 2019; Accepted: 24 April 2019; Published: 1 May 2019

Abstract: The aim of the present study was to determine how variation in the biscuit matrix affects both the degree of in vitro fragmentation and the starch hydrolysis that occurs during the oral phase of digestion. Using three different oat ingredient types (oat flour, small flakes, and big flakes) and baking powder (or none), six biscuits with different matrices were obtained. The instrumental texture (force and sound measurements) of the biscuits was analyzed. The samples were then subjected to in vitro fragmentation. The particle size distribution and in vitro oral starch hydrolysis over time of the fragmented samples were evaluated. The results showed that the samples presented different fragmentation patterns, mainly depending on the oat ingredient type, which could be related to their differences in texture. The biscuits made with oat flour were harder, had a more compact matrix and showed more irregular fragmentation and a higher percentage area of small particles than those made with big oat flakes, which were more fragile and crumbly. The highest degree of starch hydrolysis corresponded to the biscuits made with flour. Conclusions: Differences in the mechanical properties of the biscuit matrix, in this case due to differences in the oat ingredient, play a role in the in vitro fragmentation pattern of biscuits and in the oral phase of starch hydrolysis.

Keywords: particle size; in vitro oral fragmentation; oral phase of starch hydrolysis

---

1. Introduction

Oral processing during eating involves cyclic mechanical breakdown and insalivation to transform food into a bolus that is easy and safe to swallow [1]. The changes that the food undergoes during the oral phase are relevant, since they play a key role in sensory perception, as reflected by the number of existing studies dealing with the relationship between oral processing and texture perception [2] and flavor release [3]. The oral phase is also considered to have a great impact on digestion and nutrient absorption, despite its short duration compared with other phases of digestion, due to the food breakdown and insalivation that take place in the mouth [4–6].

A threshold of particle size and level of lubrication by saliva is required to form a well-lubricated, cohesive, ready-to-swallow bolus [3,7]. In starchy foods, such as biscuits, starch hydrolysis by salivary enzymes clearly begins in this in-mouth step [8] and depends on the initial structure and breakdown path of the food in the mouth. In turn, enzymatic hydrolysis of starch impacts the blood glucose response. Several studies on the impact of oral digestion have been conducted, both in vivo and in vitro [9,10].

At the same time, the number of studies of bolus properties has been increasing in recent years, providing insight about the physical and chemical changes that food products undergo during the oral phase [11]. Several methodological approaches to analyzing the properties of the bolus (obtained in vivo or in vitro) have been developed, measuring bolus particle size, bolus water content, and bolus rheology [12].

Studies of fragmentation and particle size distribution patterns in boluses provide information on the different comminution and agglomeration mechanisms that take place in the mouth during oral processing. A variety of methods to measure particle size distribution in the bolus have been applied, including sieving [11,12], laser diffraction [13,14], and image analysis [15]. The suitability of one method or another is determined by the specific characteristics of the food particles to be measured [13]. There is abundant information on the breakdown patterns and particle size distributions of boluses formed both in vivo and in vitro, using a wide variety of food products, based on studies considering the following groups of food: jelly, carrot, cheese, and nuts [16]; bread, cake, peanuts, and cheese [17]; peanuts, carrots, olives, mushrooms, egg, ham, chicken, cheese, and coconut [18]; nuts and vegetables [13]; bread and pasta [5]; cereal puffs and flakes [19]; and biscuits [20]. These studies have observed different fragmentation paths and a wide range of particle size distributions before swallowing, depending on the different characteristics of the food. The differences are related not only to the initial composition of the food products, but also to their initial structural and mechanical properties. In addition, in the case of in vivo studies, the results are also influenced by inter-individual differences in mastication [11,12,18,21,22].

The enzymes in human saliva include α-amylase. This enzyme hydrolyses starch in the mouth, which brings about the first step in starch degradation and in the digestion of starchy foods [23]. In many studies that have focused on the digestion of cereal-based food matrices, the sample has been ground, and the size and distribution of particles in an actual bolus during and after oral processing has normally been neglected. The study of oral starch hydrolysis is relevant as it influences the metabolic response when consuming starchy food and can also modify tastant release, altering the perception of saltiness, for instance [24]. The more disrupted the food matrix is, the higher the surface area available for enzymatic degradation by salivary and pancreatic amylase is, and the higher the blood glucose and/or postprandial insulin response is [8,25,26].

Little information is available on the relationship between the initial fragmentation and the degree of enzymatic degradation of cereal-based foods in the oral phase. One study of bread and spaghetti boluses obtained in vivo determined and compared the level of particle degradation and digestion before swallowing for each food [8]; the results highlighted the influence of oral digestion on the entire digestion process. In another recent study using white bread [27], different in vitro oral processing methods were applied to obtain boluses and their characteristics were compared with those of in vivo boluses; the kinetics of starch hydrolysis in the different boluses were also compared by performing in vitro gastrointestinal digestions.

Thus, it could be of interest to study in depth how the initial food matrix influences fragmentation and starch hydrolysis in the oral phase of digestion.

The present study aims to determine how variations in the biscuit matrix (not in composition), involving three different types of oat ingredients and the presence or absence of baking powder, affect the fragmentation pattern and the degree of oral starch hydrolysis in an in vitro approach.

2. Materials and Methods

2.1. Samples

Six oat biscuit formulations were prepared with three different oat ingredient types (flour, small flakes, and big flakes) and with or without baking powder (a mixture of diphosphate and sodium bicarbonate; Bakels aromatics, Gothenburg, Sweden). In all the formulations, the sum of the all-purpose wheat flour and the oat ingredient was kept constant, as was the proportion of the remaining ingredients except baking powder (vegetable oil, sugar, salt, and water).

The oat flour was sieved with a 2 mm mesh sieve and only the fraction that passed through the mesh was employed. The flake size was approximately 8 × 5 × 1 mm (length × width × thickness) for the small flakes and 11 × 6 × 1 mm for the big flakes.

The ingredients (expressed in g/100 g of a 90:10 wheat flour/oat ingredient mixture) were 10 g of sugar, 20 g of vegetable oil, 5 g of salt, 66 g of water; when applicable, 0.8 g of baking powder was also added. The oat biscuits were prepared in a single stage in a Varimixer AR (Varimixer A/S, Denmark) at 120 rpm for 2 min. The dough was sheeted and molded to 5 mm thick. After molding, the dough was cut into small discs (75 mm diameter), which were placed in the oven and baked at 200 °C for 20 min. The samples were tempered to room temperature and then stored under refrigeration (4 °C) until the tests took place. Figure 1 shows the biscuits corresponding to each formulation.

Figure 1. Photograph of the biscuits. OF: oat flour; SO: small oat flakes; BO: big oat flakes; +BP: with baking powder.

2.2. *Instrumental Texture of Samples*

A three-point bending test was performed by breaking the biscuits with a TA-XT.plus Texture Analyzer equipped with Texture Exponent 32 software (version 6.0.6, Stable Microsystems, Godalming, UK). A three-point bending rig (A/3 PB) was used and the experimental conditions were as follows: supports 50 mm apart, a probe travel distance of 50 mm, a trigger force of 0.098 N, and a test speed of 0.5 mm/s [28].

Acoustic events in decibels (dB) were registered synchronically with force. A Bruel & Kjaer free-field microphone (8-mm diameter) coupled to the texture analyzer was used. The microphone was calibrated using a Type 4231 acoustic calibrator (94 and 114 dB SPL-1000 Hz) and placed near the sample (3-cm distance, 45° angle). Ambient and mechanical noise was filtered out (1 kHz high pass filter) [28].

The data acquisition rate was 500 points/s for both force and acoustic signals. All the tests were performed at room temperature in a laboratory with no special soundproofing facilities. The maximum peak force (MF) and slope to the maximum peak force (SMF) were extracted from the force–time curves. The number of sound events up to the maximum peak force (NSP) at different thresholds (>8 and >12 dB) in each curve was obtained from the sound–time curves. Five replications were conducted for each formulation.

2.3. In Vitro Fragmentation

The biscuits were cut into 5.0 ± 0.5 g pieces (a comfortable bite size corresponding to a quarter of biscuit) and chopped using three strokes of a hand-operated mechanical food chopper (Tescoma, Cazzago San Martino, Italy). This kitchen gadget can effect cutting events that are intended to mimic the action of teeth, so the fragmentation pattern of the samples subjected to this instrumental chopping was studied as an approach to reproducing in-mouth breakdown during the first chews. A parallel in vivo preliminary test was performed for comparison purposes: 10 subjects were asked to chew 3 times on the samples (data not shown); superficial observation of the results revealed a similar fragmentation pattern.

The particles obtained by the chopping procedure were sieved using an 841 μm mesh sieve. The fraction that passed through the sieve was weighed to calculate the percentage of particles (*w/w*) smaller than 0.7 mm^2. The particle size distribution of the remaining fraction (>0.7 mm^2) was determined by image analysis. The particles were spread out on a clean, dry, transparent glass surface (30 × 21 cm), carefully separated from each other, and their image was digitized in TIF format at 600 ppi by a scanner (Canon MP270 model K10339, Lake Success, NY, USA), using a black background. The resulting images were analyzed using Nis-Elements® BR 3.2 software (Nikon, Tokyo, Japan). For this purpose, the images were binarized according to the predefined intensity threshold value, using a histogram-based segmentation process. All the objects were checked, and any unsuitable artefact (fibers and particles touching the frame) was excluded from the evaluation. The particle areas were calculated at the points where the cumulative area was 25% (a25), 50% (a50, median), and 75% (a75), as was the interquartile ratio (a75/a25). The analyses were performed in duplicate for each sample.

2.4. In Vitro Starch Oro-Digestion

The samples, cut into one-bite size pieces of approximately 5 g as described above, were placed in a perforated basket and submerged in 150 mL of simulated salivary fluid (SSF) [29]; the SSF (sodium bicarbonate, 5.208 g/L; potassium phosphate dibasic trihydrate, 1.369 g/L; sodium chloride, 0.877 g/L; potassium chloride, 0.477 g/L; calcium chloride dehydrate, 0.441 g/L; mucin from porcine stomach type II, Sigma, M2378, 2.16 g/L; α-amylase type VI-B from porcine pancreas, Sigma, A3176; 8.70 g/L, 200,000 units; distilled water) was adjusted to pH 6.9. All the reagents were of analytical grade.

The system was kept under constant agitation in order to achieve homogeneous sampling. An aliquot of 100 μL of the liquid was taken every 30 s up to a total time of 180 s. At each sampling time (30, 60, 90, 120, 150, and 180 s), the reaction in the aliquot was stopped quickly with 0.3 M sodium carbonate and ice, and the sample was frozen (−18 °C) for analysis on the following day.

The glucose released through starch degradation of the samples at the different points in time was quantified by the glucose oxidase-peroxidase (GOP) reaction, employing a K-GLUC enzymatic kit (Megazyme, Wicklow, Ireland). The free glucose was determined by treating the samples with the GOP reagent at 50 °C for 20 min, and the absorbance was read with a spectrophotometer at 510 nm (Biochrom Ultrospec 2100, Cambridge, UK). A glucose solution (1 mg/mL) was used as a standard. The analyses were performed in triplicate for each sample and sampling time. The degree of starch hydrolysis was calculated as the proportion of released glucose to sample weight.

2.5. Statistical Data Analysis

ANOVA of two factors (oat ingredient type and baking powder) and their interaction was conducted on the results of the instrumental texture analysis (maximum force, slope at maximum force, and a number of sound peaks >8 and >12 dB) and on the fragmentation parameters (percentage weight of particles with areas <0.7 mm^2 for a25, a50, a75, and the a75/a25 interquartile ratio, and a particle size in the 0.7–5, 5–50, and >50 mm^2 area ranges).

To study the effect of oat ingredient type and of baking powder or none on starch hydrolysis, an ANOVA of three factors (oat ingredient type, baking powder, and time) and their binary interaction was conducted.

The significance of the differences between mean values was determined by Fisher's least significant difference (LSD) test ($\alpha = 0.05$). These analyses were performed with XLSTAT statistical software (version 2019, Addinsoft, Paris, France).

3. Results

3.1. Instrumental Texture of Oat Biscuits

According to the ANOVA results, the interaction between the oat ingredient type and baking powder factors was not significant for any of texture parameters studied. The samples presented significant differences in instrumental texture parameters depending on the oat ingredient type ($p < 0.001$) (Table 1). Baking powder had a significant effect on NSP >12 dB ($p = 0.012$), but not on the other parameters (NSP > 8 dB, $p = 0.06$; MF, $p = 0.766$; SMF, $p = 0.791$).

Table 1. Mean values of maximum peak force (MF), slope at maximum peak force (SMF), and the number of sound peaks (NSP) above the 8 and 12 dB thresholds, measured during fragmentation of the biscuits using a three-point bending rig.

Sample [1]	MF (N)	SMF (N/s)	NSP (>8 dB)	NSP (>12 dB)
OF	$9.6 \pm 1.5^{a,b}$	7.1 ± 0.5^{a}	12 ± 3^{a}	1 ± 1^{a}
OF+BP	11.3 ± 2.0^{a}	7.6 ± 1.2^{a}	24 ± 8^{b}	6 ± 4^{b}
SO	$9.9 \pm 0.8^{a,b}$	$6.3 \pm 0.9^{a,b}$	$28 \pm 3^{b,c}$	5 ± 3^{b}
SO+BP	$8.3 \pm 0.8^{b,c}$	6.4 ± 0.9^{a}	$26 \pm 6^{b,c}$	$7 \pm 2^{b,c}$
BO	6.8 ± 2.6^{c}	$4.9 \pm 1.5^{b,c}$	$34 \pm 11^{c,d}$	$10 \pm 4^{c,d}$
BO+BP	6.3 ± 2.0^{c}	4.1 ± 1.1^{c}	41 ± 7^{d}	13 ± 2^{d}

[1] OF: oat flour; SO: small oat flakes; BO: big oat flakes; +BP: with baking powder. Mean values in the same column that do not share letters are significantly different ($\alpha = 0.05$) according to Fisher's least significant difference (LSD) test.

The big oat flake biscuits required significantly less force to break them and had lower slopes at the breaking point and a higher number of acoustic events above both sound thresholds (>8 and >12 dB) than the biscuits made with oat flour. The mechanical behavior values of the small-flake biscuits were intermediate.

The presence of baking powder significantly increased the number of sound peaks in the biscuits made with oat flour. In the samples made with the other two oat ingredient types, no significant differences were found in relation to the presence of baking powder.

3.2. Particle Size Pattern

The weight fraction of the particles with an area of less than 0.7 mm^2 was higher for the biscuits made with oat flour than for those made with flakes (average values 9.0%, 8.5%, and 7.2% for oat flour, small flake, and big flake biscuits, respectively), although these differences were not significant ($p = 0.062$). Baking powder had no significant effect ($p < 0.455$) on the weight fraction values.

The number and size distribution of particles larger than 0.7 mm^2 were analyzed through image analysis. Figure 2 shows two examples of binarized images of the fragments from different samples.

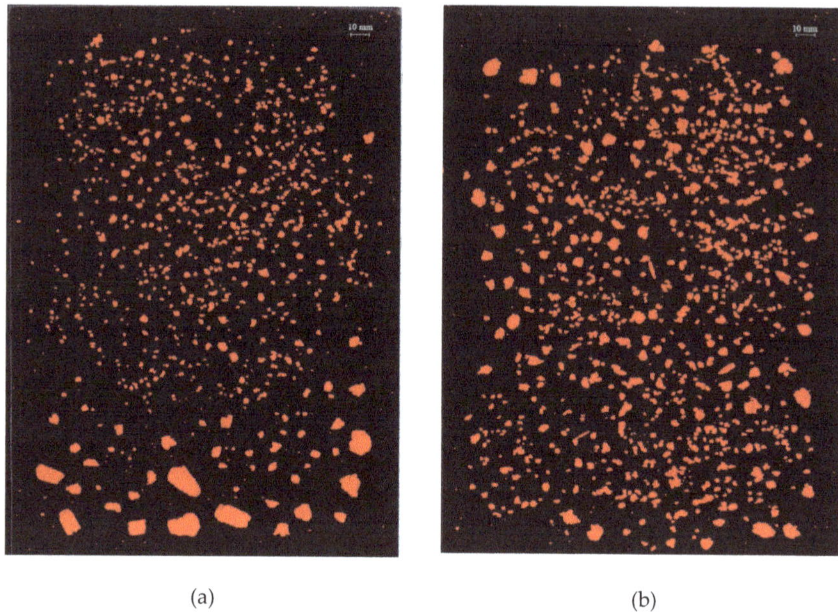

(a) (b)

Figure 2. Binarized images showing the particles generated by in vitro fragmentation of samples formulated with (**a**) oat flour with baking powder and (**b**) big oat flakes with baking powder.

Table 2 shows particle area values at 25% (a25), 50% (a50), and 75% (a75) of the cumulative area and the interquartile ratio (a75/a25).

Table 2. Particle area for 25%, 50%, and 75% of the cumulative area and the a75/a25 interquartile ratio of the biscuits after in vitro fragmentation.

Sample [1]	a25 (mm^2)	a50 (mm^2)	a75 (mm^2)a	a75/a25
OF	3.5 ± 0.7 [a]	9.4 ± 2.8 [a]	29.1 ± 7.5 [a]	8.2 ± 0.6 [a,b]
OF+BP	3.3 ± 0.1 [a]	8.9 ± 1.5 [a]	33.6 ± 6.3 [a]	10.2 ± 1.8 [a]
SO	3.8 ± 0.3 [a,b]	9.2 ± 0.4 [a]	23.2 ± 2.5 [a]	6.1 ± 0.2 [b]
SO+BP	4.1 ± 0.1 [a,b]	10.0 ± 1.1 [a]	22.8 ± 0.6 [a]	5.5 ± 0.2 [b]
BO	5.0 ± 0.6 [b,c]	10.8 ± 1.4 [a]	25.9 ± 2.7 [a]	5.2 ± 0.1 [b]
BO+BP	5.4 ± 0.8 [c]	12.1 ± 2.5 [a]	22.9 ± 0.9 [a]	4.3 ± 0.5 [b]

[1] OF: oat flour; SO: small oat flakes; BO: big oat flakes; +BP: with baking powder. Mean values in the same column that do not share letters are significantly different ($\alpha = 0.05$) according to Fisher's least significant difference (LSD) test.

Parameter a25 was the only one that presented significant differences depending on oat ingredient type ($p = 0.007$). The median particle area (a50) and a75 values did not vary significantly between biscuits made with different types of oat ingredient ($p = 0.235$ and $p = 0.071$, respectively). Baking powder had no significant effect on any of the area parameter values ($p = 0.650$, $p = 0.647$ and $p = 0.887$ for a25, a50, and a75, respectively).

The interquartile ratio (a75/a25) varied between biscuits depending on the oat ingredient type ($p = 0.001$), but no differences depending on baking powder addition were found ($p = 0.678$). The a75/a25 ratio gives information on the heterogeneity of the particle size; higher values of this parameter indicate a higher degree of heterogeneity in particle size [30]. The biscuits made with oat flour and baking powder presented a significantly higher a75/a25 ratio value than the rest, which points to a more irregular fragmentation of this biscuit (Figure 2a).

The percentage areas for each sample of particles in the different size ranges (0.7–5, 5–50, and >50 mm^2) are shown in Figure 3.

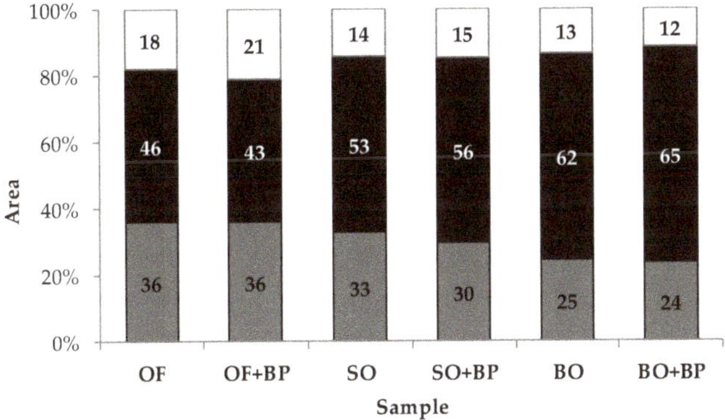

Figure 3. Percentage area corresponding to particles in the different size ranges. Grey: 0.7–5 mm^2; black: 5–50 mm^2; white: > 50mm^2. OF: oat flour; SO: small oat flakes; BO: big oat flakes; +BP: with baking powder.

The percentage areas of the small and middle-sized particle ranges varied significantly depending on the oat ingredient type. ($p = 0.008$ and $p = 0.034$, respectively). The biscuits made with big oat flakes presented a higher percentage of middle-sized particles (5–50 mm^2) than the biscuits made with oat flour, which generated a higher percentage of small particles (0.7–5 mm^2). The biscuits made with small oat flakes presented an intermediate pattern. These results indicated that the biscuits made with big flakes broke more homogeneously than those made with flour, which agrees with the a75/a25 interquartile ratio results.

3.3. *In Vitro Starch Hydrolysis*

Hydrolysis of the starch in the fragmented samples was investigated (Figure 4). As expected, the amount of glucose released increased over time for all the biscuit formulations ($p < 0.0001$). The degree of starch hydrolysis over time varied significantly depending on the oat ingredient type ($p < 0.0001$) but was not affected by the addition of baking powder ($p = 0.854$). No significant interactions between oat ingredient type and time effect were detected.

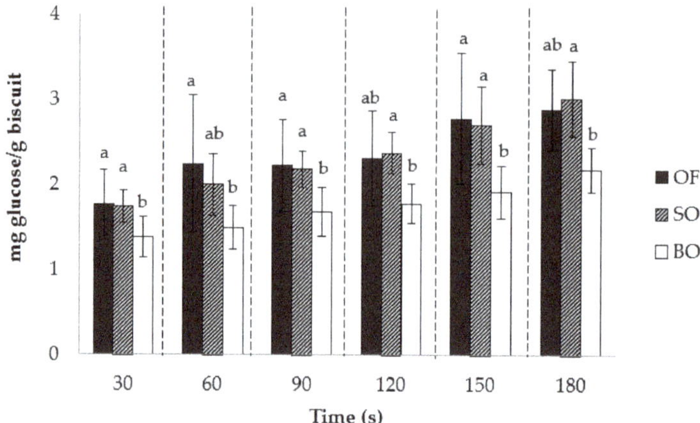

Figure 4. Mean values and error bars (standard deviation) of glucose released from samples over time. Mean values that do not share letters for the same period of time are significantly different ($\alpha = 0.05$) according to Fisher's least significant difference (LSD) test.

Overall, the biscuits made with oat flour released more glucose than those made with big flakes. This would indicate that the oat-flour biscuit matrix was more susceptible to enzymatic hydrolysis than that of the two formulations containing big flakes.

4. Discussion

4.1. Instrumental Texture of Biscuits

The differences in instrumental texture parameter values between the samples can mainly be attributed to differences in the biscuit matrix due to the oat ingredient type used.

The only effect found for baking powder was a higher number of sound peaks for biscuits made with oat flour (OF+BP) than for the same biscuit without this addition (OF), although they were equally hard at breaking point. A biscuit has a matrix composed of a flour, sugar, and fat mass in which gas cells of different sizes and shapes are imbibed [31]. It is hypothesized that baking powder increases dough aeration by creating air cells in the matrix during baking. When a biscuit breaks, a higher number of sound emission peaks would be associated with a greater number of fragile crackling sounds, which in turn would be related to the presence of a higher number of air cells in the matrix [32]. However, this increase in sound emission due to baking powder was not detected when comparing the two pairs of biscuits made with flakes (SO vs. SO+BP and BO vs. BO+BP).

In general, the big oat flake biscuits required less force to break them and produced a higher number of acoustic events. It seems that the big oat flakes make the biscuit matrix more fragile, easier to break, and more crumbly than the oat flour biscuits. The greater effect of the big particles in the biscuit matrix probably made the effect of the baking powder more difficult to detect. It has been reported that structural irregularities and anisotropy of the structure (in the present case introduced by flakes in the biscuit dough) might have a role in fracture and crack propagation through a brittle material [33] such as a biscuit. The fractures will travel quickly resulting in sudden drops in the force curve; the fracture is then somehow inhibited and stops, only to start again as the biscuit is deformed further[34]. These drops were associated with sound events.

The mechanical behavior of the small-flake biscuits was intermediate.

Auditory cues related to sensory perceptions such as crackliness, crispness, or crunchiness are important drivers of liking in a number of foods [35]; however, in the present study the sound features were measured to relate them to the breaking characteristics of the biscuits.

4.2. Particle Size Pattern

The bigger oat flake biscuits fragmented into bigger pieces than the oat flour ones, which could be associated with the differences in their mechanical properties. A fragile and easily broken matrix (due to big flakes) led to a fragmentation path with bigger particles compared to the biscuits made with oat flour, while the latter had a harder matrix that yielded small particles coexisting with big ones after fragmentation. It could be hypothesized that the inhomogeneity provided by the big oat flakes created places that were preferential for initiating breaking, since cracks would start at the weakest points and structural failure could trigger cracks in nearby structural elements [28]. Along the same lines as the results of the present study, a recent study [36] using extrudates with the addition of native or fermented rye bran concluded that it is the structural attributes of the extrudates, rather than their core composition, that dictates the breakdown pattern during mastication. Another study by the same research team [37] investigated the differences in mastication (monitored by electromyography) between breads made with three different types of rye ingredients (wholemeal, endosperm, and endosperm plus gluten) and with wheat flour. The particle size distribution of the in vivo boluses indicated that wheat bread was degraded to smaller particles than rye breads during mastication, which was related to certain physical properties of the bread, such as porosity and specific volume. The authors highlighted that the initial bread matrix played a role in the degree of in-mouth breakdown. In another study that ground bread and two kinds of pasta with a modified meat grinder, the size distribution of the mechanically fragmented samples was similar to those of the corresponding in vivo boluses [5] and the differences in the degree of fragmentation of the different samples were attributed to texture differences among them. In the same way, an in vivo investigation of cereal-based (sponge cake and brioche) food boluses in elderly people [38] found that the particle size distribution in the boluses at different mastication times depended, among other factors, on the initial structure and mechanical properties of the food products.

4.3. In Vitro Starch Oral Hydrolysis

Some research has been conducted on in vitro methods for studying the degree of in-mouth breakdown in relation to its impact on subsequent digestion. For example, a very recent study [27] using white bread compared four different in vitro oral processing methods (cutting with a knife, cutting and grinding with a pestle, blending, and grinding) to investigate the effect of fragmentation on bolus formation and on in vitro digestion. Another study [10] subjected pastas with different compositions to four different breakdown procedures (chewing by one volunteer, grinding/maceration, cutting with a knife, and cutting with a knife and grinding with a pestle) followed by in vitro gastric digestion. In these two studies, the oral digestion phase was not evaluated. Other studies [39] on oral digestion of starch have indicated the potential role of salivary α-amylase activity in the perception of cooked starch.

More abundant are studies on particle sizes and distribution in in vivo boluses and their influence on digestion, although ones in which the oral phase of digestion is studied do not abound. In vivo, ready-to-swallow wheat and rye bread boluses have been compared for particle size distribution and saliva impregnation [37]; the starch hydrolysis rate in the early phase of digestion due to the salivary amylase present in the boluses was related to the differences in the structure of the breads rather than to the size of the particles in the boluses.

Particle size in the digesta after in vivo mastication and in vitro digestion [26] has been compared in breads with different compositions and textures, made with refined or wholegrain wheat and rye flour and using either a straight dough or a sourdough process. Some other factors affect the starch oral digestion. For example, the change of cooked rice grain structure in oral digestion step [40] has been showed to influence the kinetics of starch hydrolysis, which in turn is related to increases in enzyme accessibility to rice starch; such grain-scale structural changes include grain tissue damages, normally observed during the oral digestion step. It was also found that salivary amylase had an impact on the structure of hydrolysis products of cooked starches [41], and the amount and profile of lower molecular weight fractions of these hydrolysates (that are likely the substrate for subsequent hydrolysis in the gut) were different based on cooking conditions or botanical sources of starch.

In the present study, the higher susceptibility of the oat-flour biscuit matrix to the action of saliva could be explained by the presence of a higher number of small particles, which are more easily attacked by α-amylase. In addition, oat flour is obtained by milling the whole groat, so thinner oat particles are available, which can also increase its susceptibility to hydrolysis. The more broken the matrix, the higher the surface available for enzymatic degradation by salivary amylase. In turn, this effect could be related to a higher blood glucose and/or postprandial insulin response.

5. Conclusions

In vitro fragmentation is a practical method in oral processing studies. It avoids the concomitant variability of in vivo studies regarding individual differences and difficulty to follow up starch hydrolysis level.

In the present study, relationships between the fragmentation pattern (particle size and distribution) due to the matrix differences in the experimental oat biscuits and the incipient oral starch hydrolysis were established.

The biscuits had very similar compositions but differences were found in consideration of three different types of oat ingredient (flour, small flakes, and big flakes) and the addition or omission of baking powder. The differences in the biscuit matrix were reflected in the instrumental texture features (force and sound measurements), which in turn affected the fragmentation patterns, and could be attributed more to the effect of the oat ingredient type than to that of the baking powder. The differences in particle size distribution in the fragmented biscuits influence the degree of starch hydrolysis during the oral phase, which is more intense in samples showing a higher number of small particles.

Author Contributions: A.G., A.T., and S.F. conceived the study; A.G. and Q.C.N. conducted the investigation; A.G., A.T., and P.V. contributed to the experimental design; A.G., A.T., and A.R. contributed to the data analysis and the interpretation of the results. A.G., A.R., A.T., and S.F. wrote the original manuscript.

Funding: This research was funded by the SPANISH MINISTRY OF THE ECONOMY AND COMPETITIVENESS, grant number AGL-2016-75403-R, and by the GENERALITAT VALENCIANA, grant number 2017/189 (Project: Prometeo).

Conflicts of Interest: The authors declare no conflict of interest.

References

1. Koç, H.; Vinyard, C.J.; Essick, G.K.; Foegeding, E.A. Food oral processing: Conversion of food structure to textural perception. *Annu. Rev. Food Sci. Technol.* **2013**, *4*, 237–266. [CrossRef]
2. Young, A.K.; Cheong, J.N.; Foster, K.D.; Hedderley, D.I.; Morgenstern, M.P.; James, B.J. Exploring the links between texture perception and bolus properties throughout oral processing. Part 2: Bolus mechanical and rheological properties. *J. Text. Stud.* **2016**, *47*, 474–483. [CrossRef]
3. Feron, G.; Salles, C. Food oral processing in humans: Links between physiological parameters, release of flavour stimuli and flavour perception of food. *Int. J. Food Stud.* **2018**, *7*, 1–12.
4. Gao, J.; Wang, Y.; Dong, Z.; Zhou, W. Structural and mechanical characteristics of bread and their impact on oral processing: A review. *Int. J. Food Sci. Technol.* **2018**, *53*, 858–872. [CrossRef]
5. Hoebler, C.; Devaux, M.-F.; Karinthi, A.; Belleville, C.; Barry, J.-L. Particle size of solid food after human mastication and in vitro simulation of oral breakdown. *Int. J. Food Sci. Nutr.* **2000**, *51*, 353–366. [CrossRef] [PubMed]
6. Mishra, S.; Hardacre, A.; Monro, J. Food structure and carbohydrate digestibility. In *Carbohydrates—Comprehensive Studies on Glycobiology and Glycotechnology*; Chang, C.-F., Ed.; InTech: Rijeka, Croatia, 2012; pp. 289–316.
7. Hutchings, J.B.; Lillford, P.J. The Perception of Food Texture—The Philosophy of the Breakdown Path. *J. Tex. Stud.* **1988**, *19*, 103–115. [CrossRef]

8. Hoebler, C.; Karinthi, A.; Devaux, M.-F.; Guillon, F.; Gallant, D.J.G.; Bouchet, B.; Melegary, C.; Barry, J.-L. Physical and chemical transformations of cereal food during oral digestion in human subjects. *Br. J. Nutr.* **1998**, *80*, 429–436. [CrossRef]
9. Freitas, D.; Le Feunteun, S.; Panouillé, M.; Souchon, I. The important role of salivary α-amylase in the gastric digestion of wheat bread starch. *Food Funct.* **2018**, *9*, 200–208. [CrossRef]
10. Foschia, M.; Peressini, D.; Sensidoni, A.; Brennan, M.A.; Brennan, C.S. Mastication or masceration: Does the preparation of sample affect the predictive in vitro glycemic response of pasta? *Starch* **2014**, *66*, 1096–1102. [CrossRef]
11. Devezeaux de Lavergne, M.; van de Velde, F.; Stieger, M. Bolus matters: The influence of food oral breakdown on dynamic texture perception. *Food Funct.* **2017**, *8*, 464–480. [CrossRef]
12. Panouillé, M.; Saint-Eve, A.; Souchon, I. Instrumental methods for bolus characterization during oral processing to understand food perceptions. *Curr. Opin. Food Sci.* **2016**, *9*, 42–49. [CrossRef]
13. Peyron, M.A.; Mishellany, A.; Woda, A. Particle size distribution of food boluses after mastication of six natural foods. *J. Dent. Res.* **2004**, *83*, 578–582. [CrossRef] [PubMed]
14. Gao, J.; Wong, J.X.; Lim, J.C.S.; Henry, J.; Zhou, W. Influence of bread structure on human oral processing. *J. Food Eng.* **2015**, *167*, 147–155. [CrossRef]
15. Tournier, C.; Devezeaux de Lavergne, M.; van de Velde, F.; Stieger, M.; Salles, C.; Bertrand, D. Investigation of oral gels breakdown using image analysis. *Food Hydrocoll.* **2017**, *63*, 67–76. [CrossRef]
16. Chen, J.; Khandelwal, N.; Liu, Z.; Funami, T. Influences of food hardness on the particle size distribution of food boluses. *Arch. Oral Biol.* **2013**, *58*, 293–298. [CrossRef] [PubMed]
17. Engelen, L.; Fontijn-Tekamp, A.; van der Bilt, A. The influence of product and oral characteristics on swallowing. *Arch. Oral Biol.* **2005**, *50*, 739–746. [CrossRef]
18. Jalabert-Malbos, M.L.; Mishellany-Dutour, A.; Woda, A.; Peyron, M.A. Particle size distribution in the food bolus after mastication of natural foods. *Food Qual. Pref.* **2007**, *18*, 803–812. [CrossRef]
19. Alam, S.A.; Pentikäinen, S.; Närväinen, J.; Holopainen-Mantila, U.; Poutanen, K.; Sozer, N. Effects of structural and mechanical textural properties of brittle cereal foams on mechanisms of oral breakdown. *Food Res. Int.* **2017**, *96*, 1–11. [CrossRef]
20. Rodrigues, S.A.; Young, A.K.; James, B.J.; Morgenstern, M.P. Structural changes within a biscuit bolus during mastication. *J. Tex. Stud.* **2014**, *45*, 89–96. [CrossRef]
21. Agrawal, K.R.; Lucas, P.W.; Prinz, J.F.; Bruce, I.C. Mechanical properties of foods responsible for resisting food breakdown in the human mouth. *Arch. Oral Biol.* **1997**, *42*, 1–9. [CrossRef]
22. Fiszman, S.; Tarrega, A. The dynamics of texture perception of hard solid food: A review of the contribution of the temporal dominance of sensations technique. *J. Tex. Stud.* **2018**, *49*, 202–212. [CrossRef]
23. Chen, J. Food oral processing: Mechanisms and implications of food oral destruction. *Trends Food Sci. Technol.* **2015**, *45*, 222–228. [CrossRef]
24. Ferry, A.L.S.; Mitchell, J.R.; Hort, J.; Hill, S.E.; Taylor, A.J.; Lagarrigue, S.; Vallès-Pàmies, B. In-mouth amylase activity can reduce perception of saltiness in starch-thickened foods. *J. Agric. Food Chem.* **2006**, *54*, 8869–8873. [CrossRef]
25. Read, N.W.; Welch, L.; Austen, C.J.; Barnish, C.; Bartlett, C.E.; Baxter, A.J.; Brown, G.; Compton, M.E.; Hume, K.E.; Storie, I.; Worlding, J. Swallowing food without chewing; a simple way to reduce postprandial glycaemia. *Br. J. Nutr.* **1986**, *55*, 43–47. [CrossRef]
26. Nordlund, E.; Katina, K.; Mykkänen, H.; Poutanen, K. Distinct characteristics of rye and wheat breads impact on their in vitro gastric disintegration and in vivo glucose and insulin responses. *Foods* **2016**, *5*, 24. [CrossRef]
27. Gao, J.; Lin, S.; Jin, X.; Wang, Y.; Ying, J.; Dong, Z.; Zhou, W. In vitro digestion of bread: How is it influenced by the bolus characteristics? *J. Text. Stud.*. in press. [CrossRef]
28. Chen, J.S.; Karlsson, C.; Povey, M. Acoustic envelope detector for crispness assessment of biscuits. *J. Text. Stud.* **2005**, *36*, 139–156. [CrossRef]
29. Mishellany-Dutour, A.; Peyron, M.-A.; Croze, J.; François, O.; Hartmann, C.; Alric, M.; Woda, A. Comparison of food boluses prepared in vivo and by the AM2 mastication simulator. *Food Qual. Pref.* **2011**, *22*, 326–331. [CrossRef]
30. Jourdren, S.; Panouillé, M.; Saint-Eve, A.; Déléris, I.; Forest, D.; Lejeune, P.; Souchon, I. Breakdown pathways during oral processing of different breads: Impact of crumb and crust structures. *Food Funct.* **2016**, *7*, 1446–1457. [CrossRef]

31. Sozer, N.; Cicerelli, L.; Heinio, R.-L.; Poutanen, K. Effect of wheat bran addition on in vitro starch digestibility, physico-mechanical and sensory properties of biscuits. *J. Cereal Sci.* **2014**, *60*, 105–113. [CrossRef]
32. Marzec, A.; Ziółkowski, T. Structure analysis of selected cereal products in the aspect of their acoustic properties. *Pol. J. Food Nutr. Sci.* **2007**, *57*, 89–93.
33. Varela, P.; Aguilera, J.M.; Fiszman, S. Quantification of fracture properties and microstructural features of roasted Marcona almonds by image analysis. *LWT-Food Sci. Technol.* **2008**, *41*, 10–17. [CrossRef]
34. Varela, P.; Salvador, A.; Fiszman, S. On the assessment of fracture in brittle foods: The case of roasted almonds. *Food Res. Int.* **2008**, *41*, 544–551. [CrossRef]
35. Chauvin, M.A.; Younce, F.; Ross, C.; Swanson, B. Standard scales for crispness, crackliness and crunchiness in dry and wet foods: Relationship with acoustical determinations. *J. Text. Stud.* **2008**, *39*, 345–368. [CrossRef]
36. Alam, S.A.; Pentikäinen, S.; Närväinen, J.; Katina, K.; Poutanen, K.; Sozer, N. The effect of structure and texture on the breakdown pattern during mastication and impacts on in vitro starch digestibility of high fibre rye extrudates. *Food Funct.* **2019**, in press. [CrossRef] [PubMed]
37. Pentikainen, S.; Sozer, N.; Narvainen, J.; Ylatalo, S.; Teppola, P.; Jurvelin, J.; Holopainen-Mantila, U.; Torronen, R.; Aura, A.M.; Poutanen, K. Effects of wheat and rye bread structure on mastication process and bolus properties. *Food Res. Int.* **2014**, *66*, 356–364. [CrossRef]
38. Assad-Bustillos, M.; Tournier, C.; Feron, G. Fragmentation of two soft cereal products during oral processing in the elderly: Impact of product properties and oral health status. *Food Hydrocoll.* **2019**, *91*, 153–165. [CrossRef]
39. Lapis, T.J.; Penner, M.H.; Balto, A.S.; Lim, J. Oral digestion and perception of starch: Effects of cooking, tasting time, and salivary α-amylase activity. *Chem. Senses* **2017**, *42*, 635–645. [CrossRef]
40. Tamura, M.; Okazaki, Y.; Kumagai, C.; Ogawa, Y. The importance of an oral digestion step in evaluating simulated in vitro digestibility of starch from cooked rice grain. *Food Res. Int.* **2017**, *94*, 6–12. [CrossRef]
41. Nantanga, K.K.M.; Bertoft, E.; Seetharaman, K. Structures of human salivary amylase hydrolysates from starch processed at two water concentrations. *Starch* **2013**, *65*, 637–644. [CrossRef]

Article

Characterizing the Dynamic Textural Properties of Hydrocolloids in Pureed Foods—A Comparison Between TDS and TCATA

Madhu Sharma and Lisa Duizer *

Department of Food Science, University of Guelph, Guelph, ON N1G 2W1, Canada; madhu@uoguelph.ca
* Correspondence: lduizer@uoguelph.ca

Received: 22 April 2019; Accepted: 28 May 2019; Published: 30 May 2019

Abstract: Pureed foods, a compensatory diet for dysphagia, require the incorporation of hydrocolloids in order to be swallowed safely. The effect of hydrocolloid addition on textural dynamics of pureed foods has not yet been investigated. Starch and xanthan were added to levels that allowed products to meet the criteria of the International Dysphagia Diet Standardization Initiative. Nine pureed carrot matrices made with two concentrations of starch, xanthan, and their blends were characterized for textural evolution using two dynamic sensory techniques: Temporal Dominance of Sensations (TDS) and Temporal Check-All-That-Apply (TCATA). Each test, with four replications, was conducted with 16 panelists. Results indicate that purees were divided into two groups based on sensory responses—grainy and smooth were the primary differentiating attributes for these two groups. Grainy was associated with starch-added samples, while samples with xanthan (alone and in blends) were smooth and slippery. For both groups, thickness was perceived during the first half of processing, adhesiveness in the second half of oral processing, and mouthcoating was perceived toward the end of processing. A comparison of results from these tests showed that both TDS and TCATA gave similar information about texture dynamics and product differentiation of pureed foods.

Keywords: temporal dominance of sensations (TDS); temporal check-all-that-apply (TCATA); pureed foods; carrots; starch; xanthan; oral processing; dynamic perception; International Dysphagia Diet Standardization Initiative (IDDSI)

1. Introduction

Texture modified foods, such as pureed foods, are a recommended management option for people with swallowing disorders, particularly those with oropharyngeal dysphagia [1]. Pureed foods are semisolid materials wherein the food structure has been destroyed during blending and pureeing. Liquids can be added to ensure proper consistency [2]. Hydrocolloids are also added and form an integral part of these foods. Hydrocolloids improve product consistency and cohesiveness and reduce syneresis of the product [3–5]. These improvements make the food safe to swallow for individuals with dysphagia. While providing the desired benefits, hydrocolloids will affect the rheological and sensory properties of the food [6,7]. They impact food microstructure, the breakdown of particles, force needed to deform the food during mastication, lubrication of the bolus, and mouthcoating [8,9]. Each of these properties impact food oral processing and ultimately sensory perception. Therefore, the effect of hydrocolloid addition on sensory perception of foods must be measured.

The two most commonly used hydrocolloids added to dysphagia-specific products are starch and xanthan [10,11]. At present, sensory studies examining the addition of these hydrocolloids to products for individuals with dysphagia have primarily focused on thickened fluids [12–16]. When starch or xanthan are added to liquids, they interact with salivary components, resulting in differences in sensory perceptions [3]. Starch imparts a grainy texture in thickened liquids while xanthan-thickened liquids are perceived as sticky and slippery [12,14]. There is less known about the effect of hydrocolloid addition on perception of semi-solid foods, particularly pureed foods. In one of the few published studies examining modified textured foods, xanthan-thickened pureed carrots were perceived as smooth, sticky, and slimy while starch thickened carrots were rated low in smoothness [17]. Starch addition also produced a product that was perceived to be denser than the xanthan-thickened product [18]. However, all of these sensory results are static and the dynamic perceptions of these foods have not yet been characterized. As food structure and associated properties change during oral processing, their perceptions also evolve and keep changing as the food is mixed with saliva, manipulated, and prepared to form a swallow-able bolus.

A number of sensory methods exist for measuring dynamic nature of perception. Two recent techniques are Temporal Dominance of Sensations (TDS) and Temporal Check-All-That-Apply (TCATA). TDS and TCATA differ in their approach. TDS is based on the philosophy of capturing the most dominant sensation at a particular time whereas TCATA captures all the sensations being perceived at a particular time, during oral processing. The advantages of TDS and TCATA over other dynamic sensory tests are that both are less time-consuming, require minimal training of judges, and can be used for up to 10 attributes during each evaluation [19–21]. The methodology and analysis of TDS data was first introduced in 2003 and since then it has been extensively used in almost all oral processing studies to understand the dynamics of sensory perception [22]. This has included examining flavors, tastes, and textures of semi-solid gel-based systems made with different sweeteners [23,24]. Textures of emulsion-filled gels have also been studied using this approach [25]. Using the technique of TDS, Varela et al, (2014) examined how hydrocolloid addition modulates the negative attributes in ice cream, e.g., iciness [26]. TCATA is a relatively recent technique, introduced in 2016 and has been used to examine flavors and taste in aqueous model systems with non-nutritive sweeteners [27,28]. While these two techniques provide a holistic picture of the dynamics of sensory attributes of a food as they evolve and change during oral processing, few studies have compared the results of these two techniques [29–31]. Those that have been conducted have not examined semi-solid foods.

Texture, the primary attribute responsible for flow of the bolus, is the driving factor for safety during swallowing of modified textured foods. The purpose of the current research was two-fold. The first objective was to investigate the effect of starch, xanthan, and their blends on dynamic texture perception in a hydrocolloid thickened pureed food. It is hypothesized that since the original texture of the food is destroyed during pureeing, the type and concentration of added hydrocolloid contributes significantly to textural properties. The second objective was to determine if TDS and TCATA provide similar or complementary information in terms of texture evolution and product differentiation considering the aspect of dynamics of sensory perception.

2. Materials and Methods

In this study, dynamic testing of semisolid foods was conducted using pureed food matrices made with two hydrocolloids and their blends. Pureed carrots made with starch and xanthan were studied for temporal evolution of texture attributes. Carrots, being part of orange colored vegetables, were selected as a model food because they are recommended for older adults (with or without swallowing difficulties) as a good source of fiber and Vitamin A [32,33]. The carrots (cultivar SV2384DL, Stokes Seeds, Canada) were generously donated by the Muck Crops Research Station, Guelph, Canada. The starch, Precisa Sperse 100, was provided by Ingredion Inc. (Bridgewater, NJ, USA). This is a cold water swelling modified waxy maize starch, indicative of a high amylopectin content. Xanthan, KELTROL®-521, was supplied by CP Kelco (Atlanta, GA, USA). It is an 80-mesh xanthan gum with low dusting characteristics and a high hydration efficiency. All carrots were cooked, pureed, and vacuum sealed in plastic bags and stored in a freezer (−20 °C) until required. To prepare the puree, carrots were peeled, sliced, and pressure cooked in an electric pressure cooker (Power Pressure Cooker XL, Tristar Products Inc., Model-PPC772) for 4 min (under the settings 'vegetables'), at a pressure of 70 kPa. For cooking, 80 g of water was added to 1 kg of peeled, sliced carrots. After cooking, carrots were strained and cooled for 15 min and pureed in a Robot Coupe food processor (R2NCLR, 1725 rpm) for 15 min with intermittent breaks, every 3 min, for stirring and mixing. The pureed carrots were vacuum sealed using a FoodSaver®2-in-1 Vacuum Sealer and immediately frozen (−20 °C) until further use.

The pureed carrot matrices with added hydrocolloids were prepared on each day of sensory testing. The frozen pureed carrots were defrosted for 14 hours in a refrigerator (4 °C). Prior to mixing the hydrocolloids, the temperature of the purees was equilibrated to room temperature (approximately 20 °C). Starch/xanthan were added to the puree and mixed using a whisk attachment in KitchenAid Artisan Mixer (325 W) for 80 s with intermittent breaks, after every 20 s, for scraping the puree from sides in order to have a uniform dispersion of hydrocolloid in the carrot puree. The control sample with no additives was also subjected to similar processing, in order to maintain consistency. The product was then transferred to Styrofoam cups and warmed to 55 °C in a temperature-controlled cabinet (Metro C5 Controlled Temperature) before being served for the sensory testing.

The concentrations of starch (0.8% *w*/*w*) and xanthan (0.4% *w*/*w*) added to the purees were selected such that the purees would meet International Dysphagia Diet Standardization Initiative (IDDSI) guidelines for modified textured foods. A broad outline of the four tests which were followed as part of IDDSI guidelines has been included in the Supplementary Section. Pureed foods fall under IDDSI level 4 (Table S1). Using the selected concentrations as the maximum level, a 3 by 2 full factorial design with 9 treatments (Table 1) was selected to investigate the effect of type and concentration of each hydrocolloid and their blends, on the dynamic sensory texture properties of pureed carrots.

Table 1. Concentrations of starch, xanthan, and starch/xanthan blends used while making pureed carrot matrices. (S—Starch, X—Xanthan, S/X—Blend of Starch and Xanthan).

Treatments	Sample	Starch (*w*/*w* %)	Xanthan (*w*/*w* %)
1	Control	0	0
2	S0.4	0.4	0
3	S0.8	0.8	0
4	X0.2	0	0.2
5	X0.4	0	0.4
6	S0.4/X0.2	0.4	0.2
7	S0.8/X0.2	0.8	0.2
8	S0.4/X0.4	0.4	0.4
9	S0.8/X0.4	0.8	0.4

2.1. Attribute Selection and Training

Two dynamic sensory tests were conducted on the puree treatments: Temporal Dominance of Sensations (TDS) and Temporal Check-All-That-Apply (TCATA). Both tests were conducted with same attributes. The attributes selected for dynamic texture evaluation are usually based on previous sensory studies of similar products [30,34]. A list of nine attributes was generated for this study based on the results of projective mapping and a trained panel examining textural perceptions in pureed carrot matrices [17,18]. Attributes selected include: adhesive, cohesive, dense, grainy, mouthcoating, slippery, smooth, thick, and thin.

The TDS and TCATA tests were completed by 16 panelists each and four replications of each test were conducted. All panelists (age between 20–40 years) were recruited from the University of Guelph. Selection was based on the availability and interest of panelist in the study. The TDS panel consisted of two males and 14 females and the TCATA panel consisted of three males and 13 females. Two panelists completed both tests. Tests were conducted 5 months apart.

This study was approved by the University of Guelph Research Ethics board (REB # 17-04-013). All subjects gave their informed consent for inclusion before they participated in the study. For each test, panelists were oriented for 2 days, 1 h each day. The purpose of this orientation was to familiarize all panelists with the selected attributes, the concept of dynamic testing, and the use of computer software for sample evaluation. On the first day, each attribute was defined and explained using one reference food (Table 2). This was done so that the panel developed a common understanding of attributes found in a pureed carrot matrix. It was clearly stated to the panel that there were no restrictions on the selection of the number of attributes and also the fact that the same attributes could be re-selected if perceived again. The concept of the test—TDS (the sensation/attribute that triggers the attention most) and TCATA (check/uncheck all the attributes being perceived or stopped being perceived) was exemplified by explaining how the attributes change over time while eating a hard cookie—starting from hard and crunchy, then moving to soft and sometimes pasty towards the end.

Table 2. Attribute definition with reference foods used for Temporal Dominance of Sensations (TDS) and Temporal Check-All-That-Apply (TCATA) test.

Attribute	Definition	Reference Food
Adhesive	The degree of sample that sticks to oral surfaces (like palate, tongue, teeth).	WOWBUTTER Creamy, Canada
Cohesive	Tendency of the product mass to stay together in one piece.	Gerber® Rice & Banana Baby Cereal (Add Water), Nestle, Canada
Dense	The compactness of the food product, or how solid it feels when you are orally processing it.	PHILADELPHIA Original Cream Cheese, Kraft Canada Inc.
Grainy	The presence of small, rough particles in the mouth surface.	Unsweetened Applesauce, Mott's® USA
Thick	This is related to the viscosity of the sample, how hard/easy (amount of effort needed) it is to move the sample in the mouth while orally processing it.	Nordica Smooth Plain Cottage Cheese, Gay Lea®, Canada
Thin	Product readily flows, related to viscosity of the sample (the opposite of thick).	0% Vanilla Yogurt (Mixed with water 1:1), IÖGO, Canada
Mouthcoating	The amount of film on the mouth surfaces, the coating on the oral surfaces inside the mouth.	Homestyle Mashed Potato, Betty Crocker, USA. Made as per instructions but used 35% cream and 2 Tbsp Margarine instead of milk.
Slippery	How the food slides/slips down to the back of the mouth.	Miracle Whip Regular, Kraft Canada Inc.
Smooth	Velvety feeling of the sample in your mouth. It is the absence of surface particles (the opposite of grainy).	Greek Yogurt (Vanilla Bean 5%), LIBERTÉ, Canada

Oral processing of a pureed food is short; therefore, the layout of attribute list was kept the same for all training and testing days so that the panelists were aware of attribute location on the screen. The attribute layout was similar for all the panelists and it was same for both TDS and TCATA. Testing was conducted on individual laptops and the data was collected using Compusense Cloud version 8.8 (Compusense Inc., Guelph, ON, Canada). The panelists were instructed to eat the whole sample (10 ± 1 g) served in the Styrofoam cups for sensory evaluation. The cups were coded with three-digit blinding codes. They were required to cleanse their palate with a bite of unsalted cracker and a sip of water after each sample. The samples were randomized across all panelists and across all replicates for each panelist.

2.2. *Data Analysis*

The data collection and analysis were similar for both the tests. First, TDS and TCATA product curves were constructed by plotting dominance (TDS)/absence or presence (TCATA) of each attribute by standardized time. For TDS curves, two additional lines were drawn for statistical interpretation. The first line, 'chance level' represents the selection of an attribute by random, was set at 0.11 (1/9). The second line 'significance level' represents statistical significance (95% confidence) for an attribute and was calculated as 0.18 [35]. An attribute is said to be significantly dominant at a given point in the mastication process if its proportion of selection is above this line of significance for that time period.

To examine the hypothesis that type and concentration of thickener will have an effect on dynamics of perception, difference curves, based on Fischer's Exact test (95% significance), were generated using Compusense. This was done on the non-standardized data, representing the actual oral processing time (OPT).

A product map was generated by conducting Canonical Variate Analysis (CVA), using the CVApack R-package [36]. In this two-way multivariate analysis of variance, samples were taken as fixed effect and panelists as random effect. The outcome is a product map with confidence ellipses for each sample, representing subject ellipses, where there is a 90% probability of specific sample selection by a subject. The layout of the ellipses and segment lines connecting samples give information about the degree of sample similarities and differentiation. The information on this map is a statistical test that complements the results shown on the product curves. To investigate the effect of hydrocolloid type and concentration on chewing time, oral processing times were determined and an Analysis of Variance (ANOVA—repeated measures) was conducted with samples as a fixed factor and panelists as a random factor. A Tukey's Honest Significant Difference (HSD) with a level of significance of $p < 0.05$ was performed to check the samples which differed significantly from one another.

A trajectory map, representing the oral breakdown path of each sample, was constructed by conducting a Principal Component Analysis (PCA) on the TDS and TCATA standardized data [37]. It gives information about progression of perceived attributes from the initial point of ingestion of food until it is swallowed. This was analyzed with the tempR package in R software version 3.5.1 using a data table where rows represented the dominance rate/citation proportion of each sample at 11 time-points, starting from 5%, 10%, 20%, ... , to 100% and the attributes tested in columns [38]. In this research, there were 9 columns (attributes) and 99 rows (11 time-points × 9 samples). The sensory trajectory was drawn by linking the 11 points of a particular sample from 5% (initial chew time) to 100% (final chew time), on the first two components of the PCA biplot of attributes and samples.

3. Results

3.1. Dynamic Product Curves

Sample curves were generated for both sensory tests to understand the effect of hydrocolloids in pureed carrots and to compare the two methods for the dynamic textural properties. These curves are in standardized format where the x-axis no longer represents the actual processing time but is the percentage of oral processing time. The Y-axis represents the dominance rate in TDS and citation proportion in case of TCATA curves. Dominance rate or citation proportion is the sum of all responses divided by the number of times the sample was evaluated, at a particular point of time.

3.1.1. TDS Curves

TDS curves are shown in Figures 1 and 2. The control sample (with no added hydrocolloid) had three dominant attributes during the entire oral processing time—grainy, thick, and mouthcoating (Figure 1A). This was similar for S0.4 (Figure 1B) but there were two more attributes perceived as dominant—dense and adhesive—at very low dominance rates. These were very close to the significance limit of 0.18. For both the samples, grainy had the highest dominance rate for the initial one-third of the OPT at which point thick became dominant until almost the end of the mastication time. Mouthcoating became dominant at approximately 80% of the mastication time. For S0.8 (Figure 1C), the dominant attributes were same as in the control sample with an additional attribute of adhesive being above the significance limit for a small duration during the mid-mastication time (45%–60% OPT).

Pureed carrots matrices made with xanthan showed a different set of attributes as dominant than the control sample. Thick and smooth were the dominant attributes for X0.2 for the majority of the OPT (Figure 1D). Slippery was dominant after the initial one-third chewing time to the end. Similar to the starch samples, mouthcoating was dominant after 80% OPT. Grainy was dominant for a very short time in the first quarter. Similar to the profile of X0.2, pureed carrots made with X0.4 (Figure 1E) was perceived as smooth for most part of chewing time. The other dominant attributes above the significance limit were thick, slippery, adhesive, and mouthcoating. Adhesive was dominant in the third quarter, while slippery and mouthcoating became dominant in the last quarter of OPT and remained that way to the end.

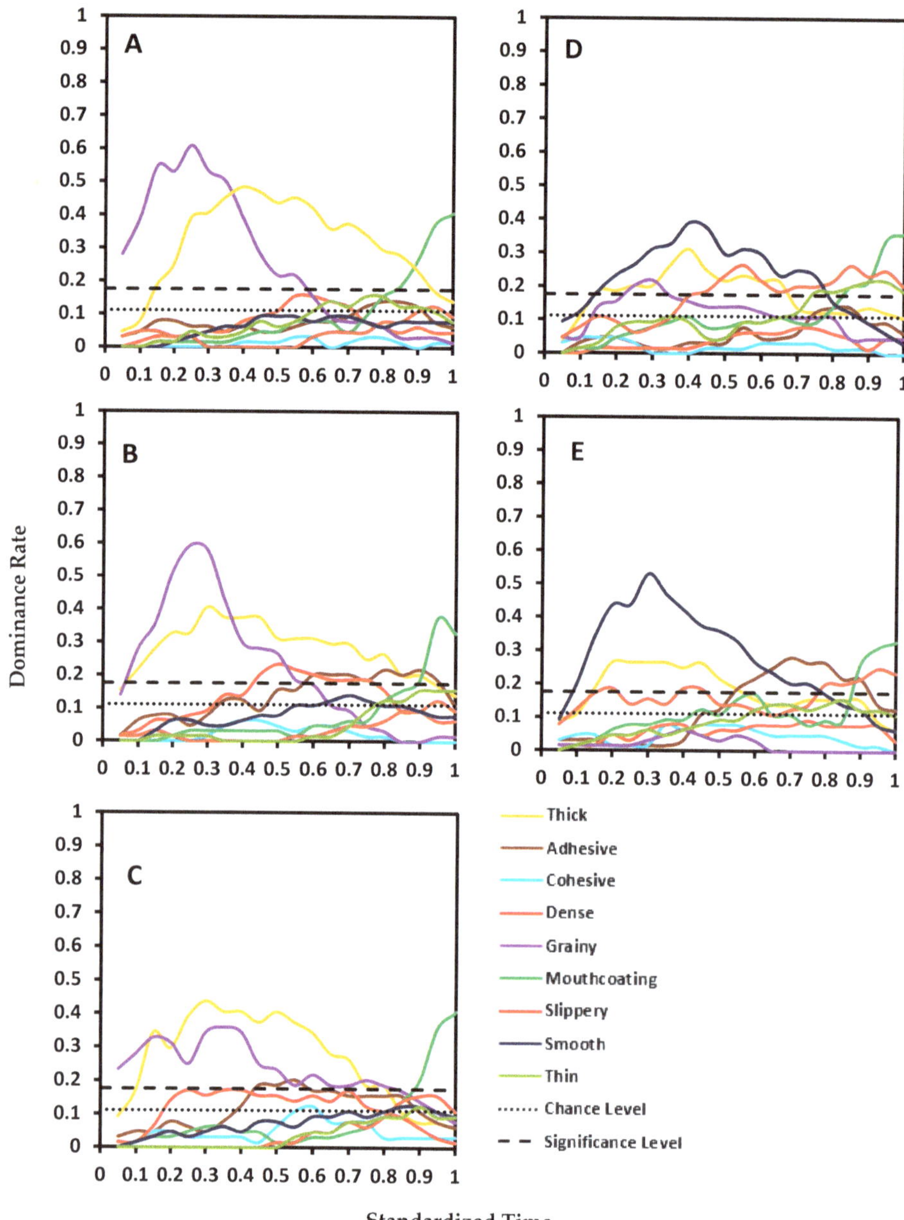

Figure 1. Standardized TDS curves of five pureed carrot matrices made with starch and xanthan: (**A**) control (no starch/xanthan), (**B**) 0.4% starch, (**C**) 0.8% starch, (**D**) 0.2% xanthan, and (**E**) 0.4% xanthan.

Each of the four blends (Figure 2A–D) had two similar dominant attributes in comparison to the control sample; thick and mouthcoating. The main difference was observed for grainy and smooth. The trend for these was opposite in the blends in comparison to the control. While grainy was dominant and smooth was below the chance limit for the control, in the case of starch–xanthan blends, smooth was a primary dominant attribute and grainy was below the chance limit. The other dominant attributes in all blends were slippery and adhesive.

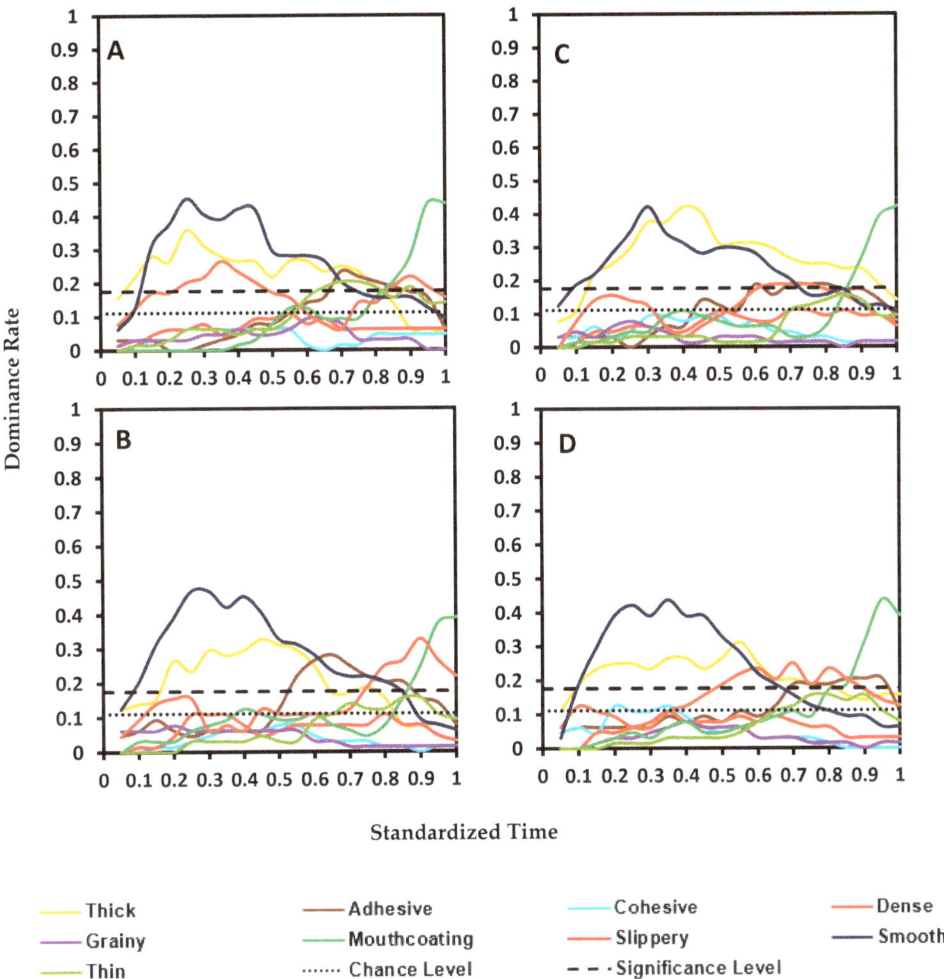

Figure 2. Standardized TDS curves of four pureed carrot matrices made with starch and xanthan blends (S—starch, X—xanthan, number represent the w/w percent concentration in pureed carrots): (**A**)—S0.4/X0.2, (**B**)—S0.4/X0.4, (**C**)—S0.8/X0.2, and (**D**)—S0.8/X0.4.

3.1.2. TCATA Curves

The dynamic TCATA curves for the 9 pureed carrot samples are shown in Figures 3 and 4. The control sample had a very high citation proportion for grainy followed by thick (Figure 3A). Grainy was the perceived attribute with highest citation proportions until the end while thick remained high until the third quarter of the OPT. The citation proportion for three attributes—adhesive, mouthcoating, and slippery were higher than thick after 75% of the mastication time. The remaining four attributes of cohesive, dense, thin and smooth had very low citation proportions. Pureed carrots matrices made with S0.4 and S0.8 showed similar trends in dynamic perception as the control sample.

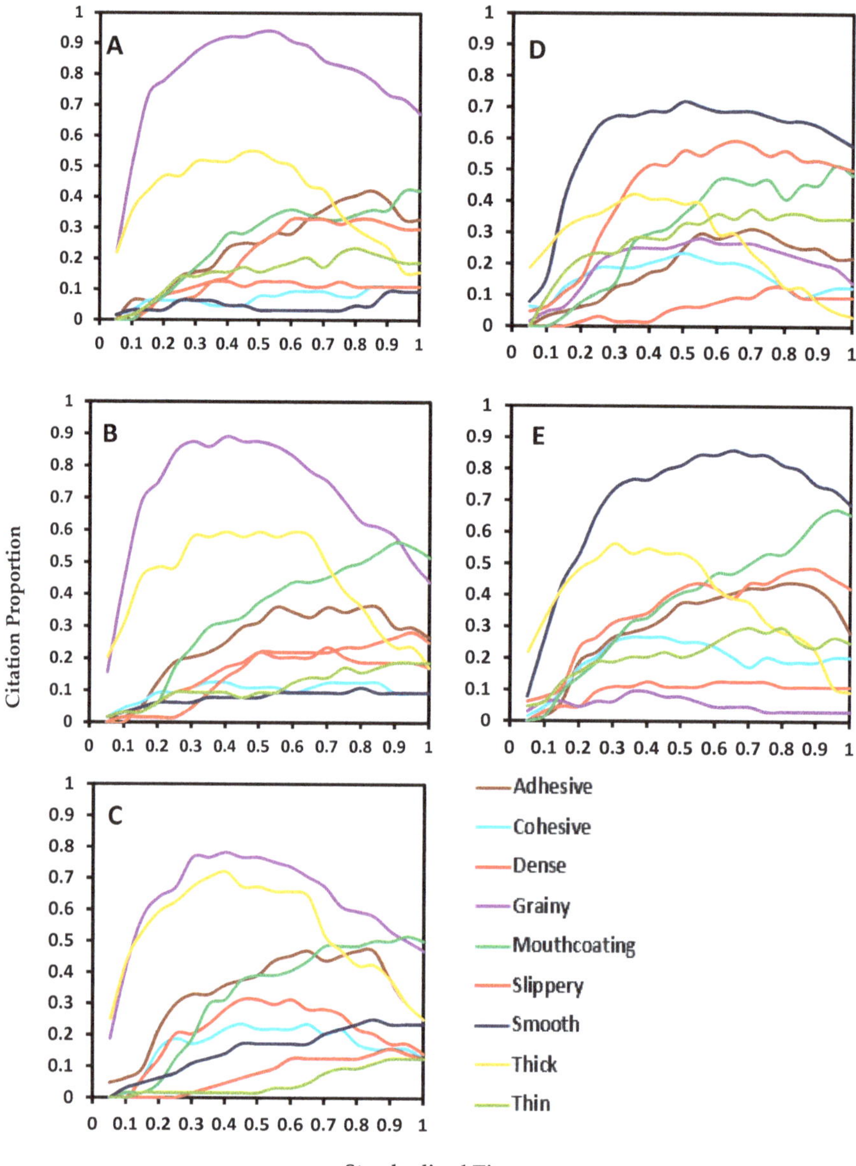

Figure 3. Standardized TCATA curves of five pureed carrot matrices made with starch and xanthan: (**A**)—control (no starch/xanthan), (**B**)—0.4% starch, (**C**)—0.8% starch, (**D**)—0.2% xanthan, and (**E**)—0.4% xanthan.

The main difference between both xanthan samples and the control sample was grainy perception. It was very low for X0.2 while it was very high in control sample for most of the mastication time. Xanthan attributed the perception of smooth in the pureed carrot matrix. Slippery and thick were the other two most cited attributes in the first half of the OPT. As the citations decreased for thick, they increased for the attributes of mouthcoating and thin.

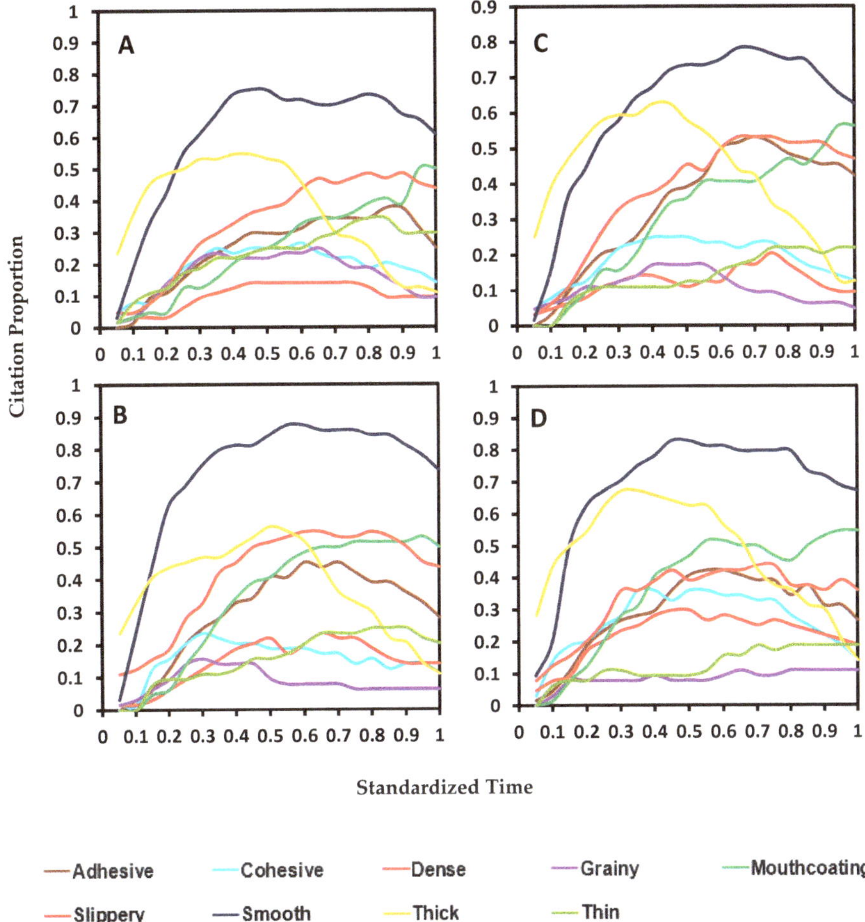

Figure 4. Standardized TCATA curves of four pureed carrot matrices made with starch and xanthan blends (S—starch, X—xanthan, number represent the w/w percent concentration in pureed carrots): (**A**)—S0.4/X0.2, (**B**)—S0.4/X0.4, (**C**)—S0.8/X0.2, and (**D**)—S0.8/X0.4.

Comparison of the blend samples (Figure 4A–D) with the control sample indicated that two attributes, grainy and smooth, had opposite trends in these samples. Grainy was very low for all blends and smooth was high for most of the chewing time. Thick was the only attribute with a similar trend and citation proportion while all the other attributes were perceived at a higher citation than in control.

3.2. Difference Curves

The difference curves have been plotted in the unstandardized format with x-axis representing the actual processing times. Difference curves for xanthan and starch products in comparison to the control sample are shown in Figures 5 and 6. As evident from these curves (Figure 5), starch samples (S0.4 and S0.8) had very few differences compared to the control for both TDS and TCATA. The few evident differences are present at a low significance level and for a very low duration. The difference attributes for both the xanthan samples against the control sample are similar with the fact that the difference rate of each different attribute increases with the concentration of xanthan. In general, xanthan samples are more smooth, slippery and have mouthcoating while the control is significantly different for grainy and thick (Figure 6). Some attributes are captured by individual tests such as mouthcoating and thick with TDS, and cohesive with TCATA. The difference curves in TCATA have longer durations for an attribute compared to the TDS because of the nature of the test.

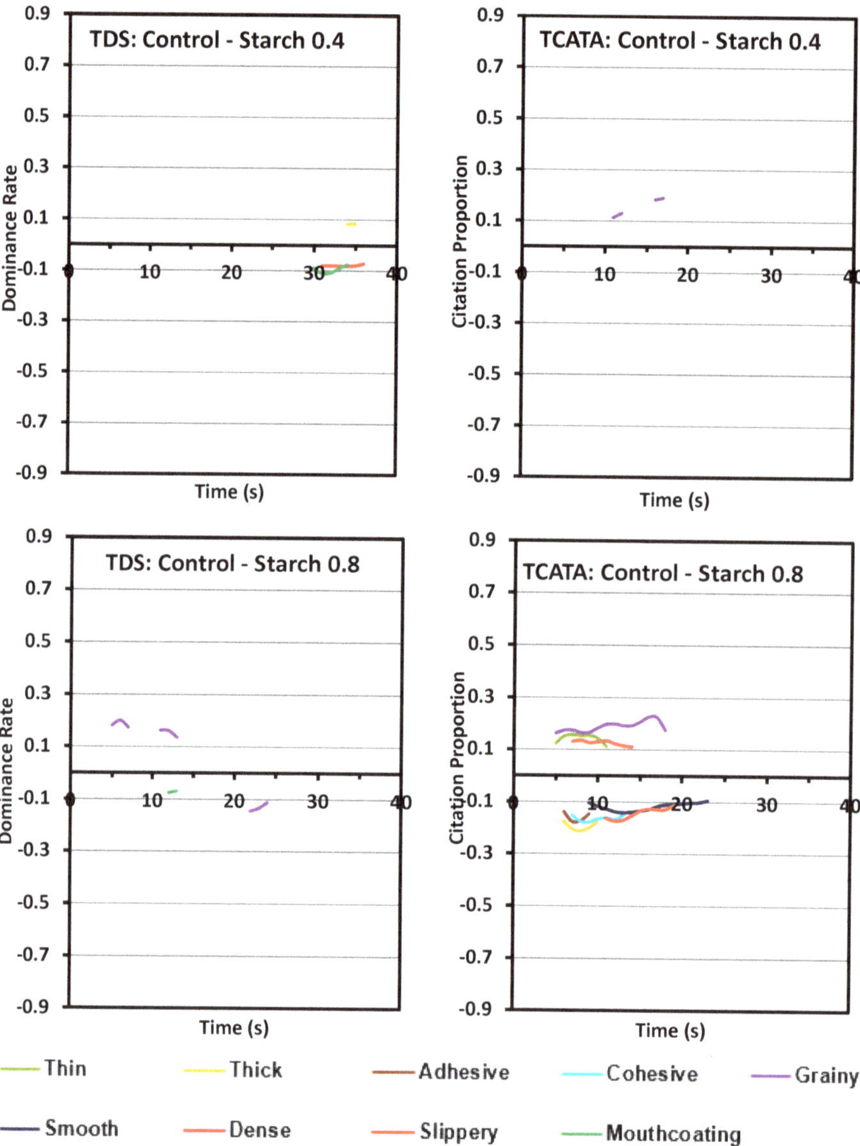

Figure 5. Difference curves of pureed carrot matrices between control and two starch concentrations (0.4% and 0.8% *w/w*). TDS (left columns) and TCATA (right columns). Comparisons are made between a pair of samples, the first and second sample are depicted respectively above and below the zero line. Significantly different attributes varying between the samples are shown, calculated at 95% Fischer's Exact test.

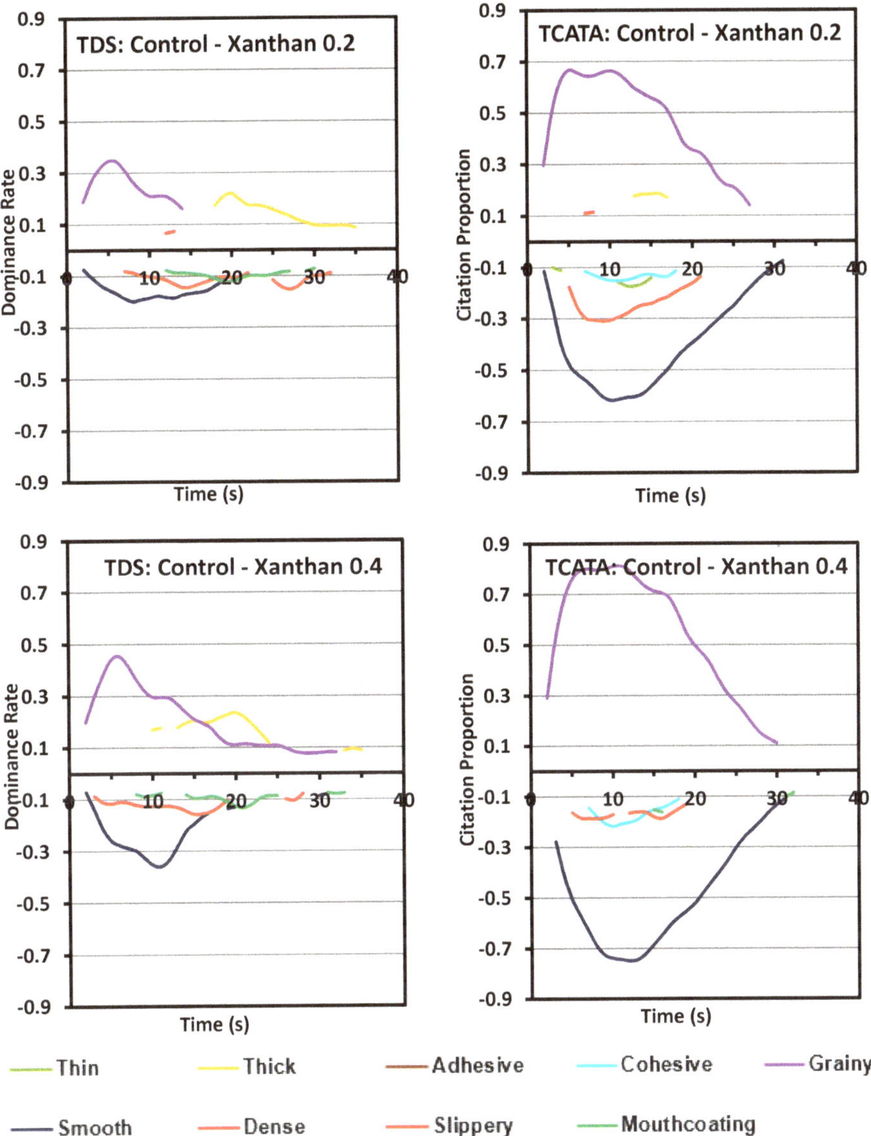

Figure 6. Difference curves of pureed carrot matrices between control and two xanthan concentrations (0.2% and 0.4% *w*/*w*). TDS (left columns) and TCATA (right columns). Comparisons are made between a pair of samples, the first and second sample are depicted respectively above and below the zero line. Significantly different attributes varying between the samples are shown, calculated at 95% Fischer's Exact test.

The similarity in texture perception between control and starch containing pureed carrot matrices is also evident from the difference curves between starch and xanthan samples. Most of the significant attributes between starch and xanthan were similar to those between control and xanthan. The differences in texture dynamics due to different concentrations of hydrocolloids used in this study were negligible in both starch and xanthan.

Within the blend samples xanthan was more influential and impacted perceptions more than starch, as shown in the Supplementary Section (Figure S1). Although this is only one example, all blends followed a similar pattern. The differences between blends was also very minimal and comparison between two blends is shown in the Supplementary Section (Figure S2).

3.3. Product Map

The product map was obtained by performing a CVA on perception duration of each attribute of the sample by panelist mean score. This biplot gives information regarding product discrimination, where the product means are separated while keeping the panelist scores as close as possible for a specific product mean [36].

As indicated in Figure 7A, the products were significantly discriminated with $p < 0.001$ and F-statistic of 5.46, for TDS. The first dimension explains a significant amount of variability (84%) of the sample discrimination. The control and both starch samples are characterized by grainy, X0.2 by thin and slippery while most of the others are characterized by smooth, slippery and thin. All blends have textural similarities with X0.4. Since mouthcoating and thick were present in all samples in approximately the same dominance rates, they are either very close to the origin or have a short attribute arrow. In other words, these two do not play a significant role in discriminating the samples.

CVA (90.97%)

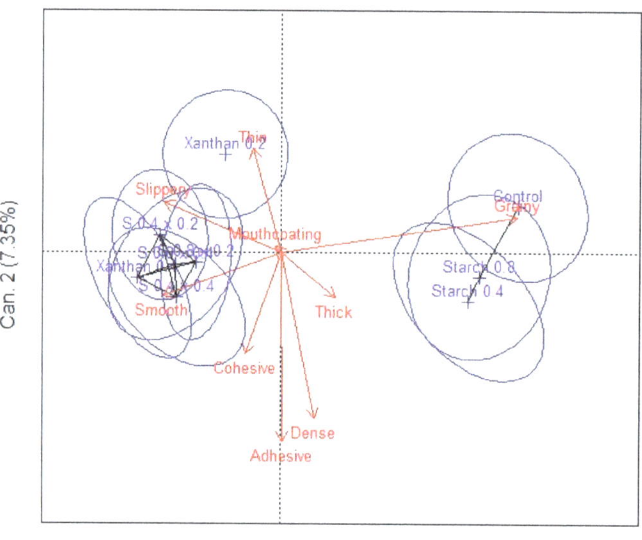

Can. 1 (83.62%)

*NDIMSIG=1, Hotelling Lawley stat=3.559, F=5.462 (p=<0.001***
Confidence ellipses=90%

(**A**) Product map for TDS

Figure 7. *Cont.*

CVA (95.26%)

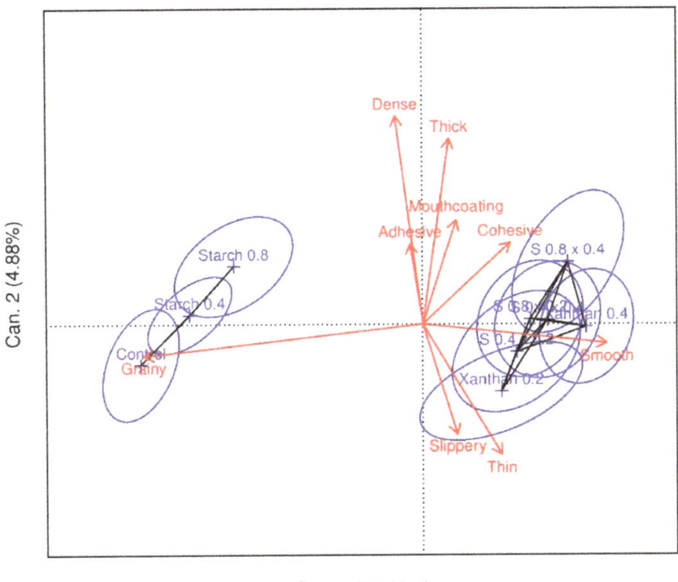

*NDIMSIG=2, Hotelling Lawley stat=7.5, F=11.511 (p=<0.001***)*
Confidence ellipses=90%

(**B**) Product map for TCATA

Figure 7. Product map discriminating the nine pureed carrot matrices (*w*/*w*) made with starch, xanthan and their blends, using CVA analysis. (**A**) TDS, (**B**) TCATA. The sample names are in blue, attributes are in red and the connecting black line between the samples is an indicator of their similarities. The samples are control (no starch/xanthan), starch (0.4, 0.8), xanthan (0.2, 0.4), S0.4/X0.2, S0.4/X0.4, S0.8/X0.2, and S0.8/X0.4.

For the CVA product-attribute biplot of TCATA (Figure 7B), 95% of sample discrimination is accounted by the first two dimensions. The first axis accounting for 90% of the data variability is primarily based on geometrical characteristics of texture—smooth and grainy. The samples are significantly discriminated with $p < 0.001$ and F-statistic of 11.51. As in the case of TDS, control and two starch samples have textural similarities. The remaining samples are closely related in texture perception, being characterized by smooth, slippery, and thin. This is very similar to TDS except for one minor thing. The connecting black line segments in the product maps, between the samples is an indicator of their similarities. According to the TDS output, the four blends are closely related to only X0.4 while, as per the TCATA output, these share similarities with both the xanthan samples. In both tests, the discriminating attributes ae grainy, smooth, slippery, and thin.

3.4. Oral Processing Time

The oral processing times were analyzed to see if addition of hydrocolloid and its concentration had an effect on the time to process the purees. No significant differences were found between samples for the oral processing times either by TDS or TCATA. The oral processing times were around 7 s more while performing TDS (range from 30.1–31.7 s) compared to TCATA (22.6–24.2 s). This difference in time between the two techniques could be an artifact of testing. Panelists have to focus on one attribute catching their attention in TDS while in TCATA they have to select all the attributes being perceived, so less focus is needed during the latter sensory evaluation.

3.5. Sensory Trajectory

The sensory trajectories of dominant rates for the nine samples, in TDS, are plotted on the PCA biplot (Figure 8). The first two dimensions explain 68% of the variance. Both the dimensions are guided strongly by geometrical textural parameters, grainy being on the first dimension and smooth on the second dimension. Mouthcoating is the other attribute contributing significantly to the first dimension. All the samples have mouthcoating as the last perceived dominant attribute. The sensory trajectories for the nine samples can be classified into two groups based on their initial perceived attribute and the pathway to reach the end-point of mouthcoating. The first group consisting of control and both starch samples start from grainy perception, moves towards thick, dense, and adhesive, and then ends at mouthcoating. The second group consisting of rest of the six samples follow the path of dense, cohesive, smooth, adhesive, thin, and mouthcoating.

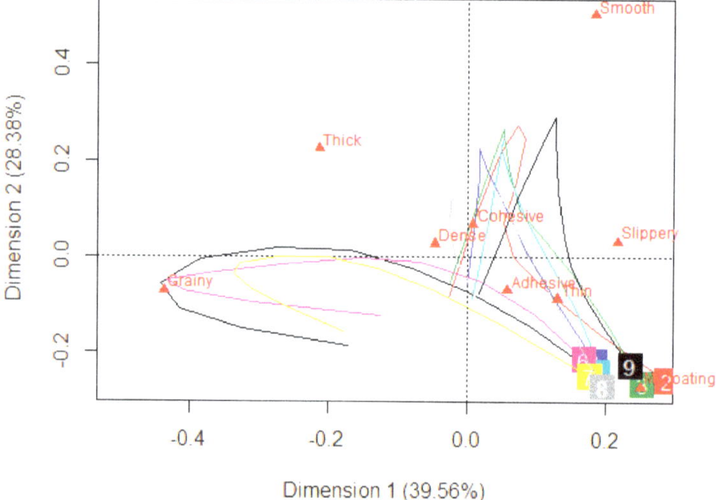

Figure 8. A PCA biplot, depicting the sensory trajectory (TDS) at 11 equally spaced time points during oral processing of nine pureed carrot matrices made with starch (S), xanthan (X), and their blends (S/X). The nine samples are each indicated by a different number and each trajectory ends where the number appears in a colored square box. Some of the numbers are overlapping due to similarities. S—starch, X—xanthan. 1-Control (black), 2-S 0.4/X 0.2 (red), 3-S 0.4/X 0.4 (green), 4-S 0.8/X 0.2 (blue), 5- S 0.8/X 0.4 (turquoise), 6-S 0.4 (pink), 7-S 0.8 (yellow), 8-X 0.2 (grey), and 9-X 0.4 (navy blue).

The sensory trajectories of citation proportions for TCATA data, represented on the first two dimensions of PCA biplot explain 82% of the variance (Figure 9). As observed in TDS, both the dimensions in this case are also strongly influenced by geometrical texture characteristics of smooth (first dimension) and grainy (second dimension). All the other attributes were placed close to each other, specifically thick, dense, thin, and cohesive. The two classification groups of all samples are evident as in the case of TDS with similar samples being in each group. The pathway being followed, showing the progression of attributes from the start to the end is not as clearly marked in the TCATA plot when compared to the TDS plot. All the samples had a common final attribute of mouthcoating in TDS which was not the case in TCATA. This is probably because of the nature of TCATA, which is based on the concept of selecting all the perceived/multiple attributes at a particular time point.

Although it could be visualized that the group consisting of control and the two starch samples were mainly in the quadrant for grainy and thick and the other samples had proximity to smooth, thin, cohesive, and slippery perception.

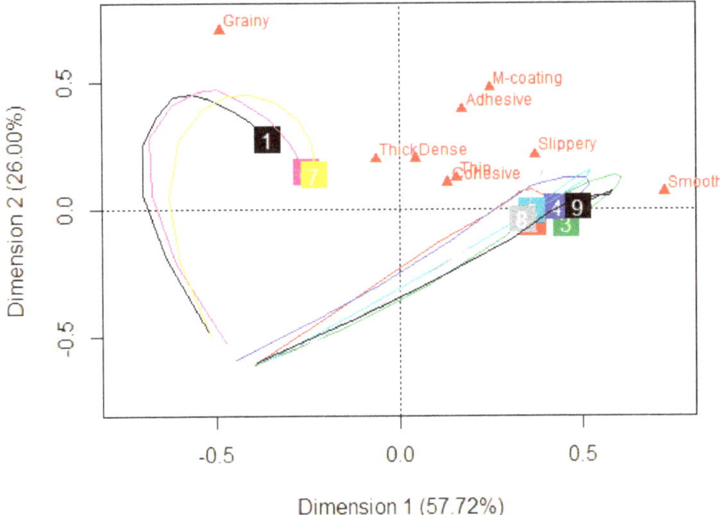

Figure 9. A PCA biplot, depicting the sensory trajectory (TCATA) at 11 equally spaced time points during oral processing of nine pureed carrot matrices made with starch (S), xanthan (X) and their blends (S/X). The nine samples are each indicated by a different number and each trajectory ends where the number appears in a colored square box. Some of the numbers are overlapping due to similarities. S—starch, X—xanthan. 1-Control (black), 2-S0.4/X 0.2 (red), 3-S 0.4/X 0.4 (green), 4-S 0.8/X 0.2 (blue), 5-S 0.8/X 0.4 (turquoise), 6-S 0.4 (pink), 7-S 0.8 (yellow), 8-X 0.2 (grey), and 9-X 0.4 (navy blue).

4. Discussion

Most published literature investigating the effect of hydrocolloids on sensory properties of semi-solid food systems have looked at starch, fat, and/or protein-based foods such as mashed potatoes, custard, and mayonnaise and have used static methodologies, with a trained panel [39,40]. Our goal was to examine the effect of commonly used hydrocolloids, starch and xanthan, on dynamic textural properties of a carbohydrate-based modified textured food system, as perceived by an untrained panel.

Pureed carrot matrices were made with two concentrations each of starch, xanthan, and four combinations of the starch–xanthan blend. These were compared to the control puree that did not contain any hydrocolloid. The dynamic texture perception of starch added pureed carrots was very similar to the control sample, with three dominant attributes—grainy, thick, and mouthcoating. The duration and the rate of grainy perception decreased with the increase of starch concentration. Addition of xanthan masked the grainy perception and reduced the perception of thick. Xanthan samples were perceived as smooth for most of the processing time followed by slippery, thin, and mouthcoating. Sensory characterization of thickened fluids highlighted similar results where starch had a higher grainy perception while xanthan attributed slipperiness [14]. The smooth perception in pureed carrots attributed by xanthan was similar to that perceived in mashed potatoes where addition of xanthan reduced the perception of granularity and enhanced creaminess [39]. This was attributed to the film being formed by xanthan around the leached amylose. The slippery perception can be associated with increased extensional flow properties of food because of the shear thinning behavior of xanthan, its interaction with salivary mucins and resistance to amylase action [3,7,41]. Probably, the same argument can be given for the xanthan samples being rated low for the perception of thick compared to the control and starch samples.

Synergistic interactions of binary hydrocolloids in the form of blends have been used in various food applications as the desired functional properties can be achieved with a reduced overall concentration of sole hydrocolloid [42,43]. In this research, the four blends of starch and xanthan tested for dynamic texture perceptions were very similar to each other and with the two xanthan samples. Even at low concentrations, xanthan masks the grainy perception of starch as observed in the blend where xanthan was added at one-fourth of starch concentration (S0.8/X0.2).

Addition of starch and xanthan increased the cohesive perception of pureed carrots, fulfilling an important criterion of adding these to the compensatory diets of people with swallowing disorders. Starch and xanthan had no effect on mouthcoating, which was similar in all samples occurring towards the end of the OPT and with almost similar dominance/citation rates. Another highlight of this study is the fact that even when the overall concentration of starch and xanthan in pureed carrots is the highest (S0.8/X0.4), dense is not a dominant attribute.

Since modified texture foods are used as a dietary intervention for people with dysphagia, an important attribute that needs to be examined is ease of swallowing. Cohesiveness and ease of swallowing are interrelated; a cohesive sample is easier to manipulate and swallow [3]. Given the nature of dynamic testing, it was not possible to include ease of swallow as an attribute, however, it should be included in future studies of food oral processing of pureed products.

TDS and TCATA product and difference curves showed similarities in attribute appearance in all samples. The major difference between the two tests were the increased duration and citation rate of attributes in TCATA compared to TDS. This is a function of test methodology. These tests also differentiated the samples in the same way with similar differentiating attributes across the nine samples, as per the CVA results (Figure 7). This is contrary to the findings of other published literature comparing TDS and TCATA. The results of a dynamic sensory study conducted on a range of products indicated better sensory discrimination with TCATA than TDS [31]. Others have concluded that, although the results were similar across TDS, TCATA, and Progressive Profiling, TCATA was felt to provide a complete product description [30]. Similarly, it was concluded that product discrimination was found to be better with TCATA and M-TDS (TDS by one modality such as texture or flavor) than TDS, when studying yogurts varying in flavor and texture [29].

An advantage of using dynamic sensory tests over static tests is the ability to visualize the sensory trajectory which provides information about the succession of perceived attributes over the oral processing time. Others have also shown this in semi-solid foods, gels, and cheese [44,45]. In the current study, the sensory trajectory generated for TCATA was less obvious than the trajectory generated from the TDS data, although the sample groupings was similar in both tests. A clear attribute path was drawn, in TDS, for the movement of a sample starting from the initial attribute to the time it was swallowed as only one attribute was selected at a point of evaluation, unlike TCATA, where multiple attributes can be selected at a particular point of time.

While this study has shown the differences in dynamic perceptions of starch and xanthan-based products, one limitation is that the attribute list was not randomized across the panelists. As suggested by Pineau et al [21], the attribute order can create a small bias on the time of selection where attributes at the top are selected before the ones at the bottom of the list. However, in the current study, attributes were listed in a 3x3 matrix with thick, thin, and adhesive at the top of the matrix. Attributes that were most often considered dominant/perceived by participants were smooth, mouthcoating, and grainy, indicating that a bias toward the attributes at the top of the list was not evident. Another limitation is that the subjects performing the tests were not the population for which modified textured foods exist. Participants used in this study were selected from a healthy population of university students and while this may have biased the results, collecting dynamic perceptions using individuals with dysphagia may not be possible to due cognitive difficulties and other health concerns. To date, no one has studied dynamic sensory perceptions with this population.

5. Conclusions

The purpose of this research was first, to examine the effect of hydrocolloid addition on sensory properties of pureed carrots. The concentration of hydrocolloid added to the carrots was less influential on dynamic sensory properties than the type of hydrocolloid used. Corn starch addition contributed a grainy texture while xanthan contributed a smoothness to the product. When blending the two hydrocolloids, xanthan was more influential than corn starch on impacting sensory perception. The second objective was to compare results between two dynamic tests. Based on the results observed in this research, both TDS and TCATA capture similar information indicating that either of the two tests can be used to examine the dynamic nature of sensory perception of a modified texture food such as pureed foods.

Supplementary Materials: The following are available online at http://www.mdpi.com/2304-8158/8/6/184/s1, Table S1: IDDSI guidelines for modified textured foods, Figure S1: Difference curves TDS (left columns) and TCATA (right columns). Comparisons are made between a pair of samples, the first and second sample are depicted respectively above and below the zero line. Significantly different attributes varying between the samples are shown, calculated at 95% Fischer's Exact test, Figure S2: Difference curves TDS (left columns) and TCATA (right columns). Comparisons are made between a pair of samples, the first and second sample are depicted respectively above and below the zero line. Significantly different attributes varying between the samples are shown, calculated at 95% Fischer's Exact test.

Author Contributions: Conceptualization, M.S. and L.D.; Data curation, M.S.; Formal analysis, M.S.; Funding acquisition, L.D.; Writing—original draft, M.S.; Writing—review and editing, L.D.

Funding: This research was supported by Natural Sciences and Engineering Research Council of Canada (NSERC) Discovery Grant (RG), grant number 400480.

Acknowledgments: Carrots were donated by the Muck Crops Research Station, Guelph, Canada. Starch was generously supplied by Ingredion Inc. and xanthan by CP Kelco.

Conflicts of Interest: The authors declare no conflict of interest.

References

1. Sukkar, S.G.; Maggi, N.; Travalca Cupillo, B.; Ruggiero, C. Optimizing Texture Modified Foods for Oro-pharyngeal Dysphagia: A Difficult but Possible Target? *Front. Nutr.* **2018**, *5*, 1–10. [CrossRef]
2. Keller, H.; Chambers, L.; Niezgoda, H.; Duizer, L. Issues associated with the use of modified texture foods. *J. Nutr. Health Aging* **2012**, *16*, 195–200. [CrossRef] [PubMed]
3. Nishinari, K.; Turcanu, M.; Nakauma, M.; Fang, Y. Role of fluid cohesiveness in safe swallowing. *NPJ Sci. Food* **2019**, *3*, 1–13. [CrossRef]
4. Funami, T. Next target for food hydrocolloid studies: Texture design of foods using hydrocolloid technology. *Food Hydrocoll.* **2011**, *25*, 1904–1914. [CrossRef]
5. Funami, T. The Formulation Design of Elderly Special Diets: Formulation design for elderly. *J. Texture Stud.* **2016**, *47*, 313–322. [CrossRef]
6. Agudelo, A.; Varela, P.; Fiszman, S. Methods for a deeper understanding of the sensory perception of fruit fillings. *Food Hydrocoll.* **2015**, *46*, 160–171. [CrossRef]
7. Abu Zarim, N.; Zainul Abidin, S.; Ariffin, F. Rheological studies on the effect of different thickeners in texture-modified chicken rendang for individuals with dysphagia. *J. Food Sci. Technol.* **2018**, *55*, 4522–4529. [CrossRef]
8. Chen, L. Emulsifiers as food texture modifiers. In *Modifying Food Texture Volume 1: Novel Ingredients and Processing Techniques*; Chen, J., Rosenthal, A., Eds.; Woodhead Publishing Series in Food Science, Technology and Nutrition; Woodhead Publishing: Sawston, Cambridge, UK, 2015; pp. 27–49.
9. Van Vliet, T. Colloidal aspects of texture perception. *Food Eng. Ingred.* **2009**, *150*, 27–40. [CrossRef]
10. Newman, R.; Vilardell, N.; Clavé, P.; Speyer, R. Effect of Bolus Viscosity on the Safety and Efficacy of Swallowing and the Kinematics of the Swallow Response in Patients with Oropharyngeal Dysphagia: White Paper by the European Society for Swallowing Disorders (ESSD). *Dysphagia* **2016**, *31*, 232–249. [CrossRef]
11. Hanson, D.B. A review of diet standardisation and bolus rheology in the management of dysphagia. *Curr. Opin. Otolaryngol. Head Neck Surg.* **2016**, *24*, 183–190. [CrossRef]

12. Matta, Z.; Chambers, E.; Garcia, J.M.; Helverson, J.M. Sensory Characteristics of Beverages Prepared with Commercial Thickeners Used for Dysphagia Diets. *J. Am. Diet. Assoc.* **2006**, *106*, 1049–1054. [CrossRef] [PubMed]
13. Vilardell, N.; Rofes, L.; Arreola, V.; Speyer, R.; Clavé, P. A Comparative Study Between Modified Starch and Xanthan Gum Thickeners in Post-Stroke Oropharyngeal Dysphagia. *Dysphagia* **2016**, *31*, 169–179. [CrossRef]
14. Ong, J.J.-X.; Steele, C.M.; Duizer, L.M. Sensory characteristics of liquids thickened with commercial thickeners to levels specified in the International Dysphagia Diet Standardization Initiative (IDDSI) framework. *Food Hydrocoll.* **2018**, *79*, 208–217. [CrossRef] [PubMed]
15. Garin, N.; De Pourcq, J.T.; Martín-Venegas, R.; Cardona, D.; Gich, I.; Mangues, M.A. Viscosity Differences Between Thickened Beverages Suitable for Elderly Patients with Dysphagia. *Dysphagia* **2014**, *29*, 483–488. [CrossRef] [PubMed]
16. Vickers, Z.; Damodhar, H.; Grummer, C.; Mendenhall, H.; Banaszynski, K.; Hartel, R.; Hind, J.; Joyce, A.; Kaufman, A.; Robbins, J. Relationships Among Rheological, Sensory Texture, and Swallowing Pressure Measurements of Hydrocolloid-Thickened Fluids. *Dysphagia* **2015**, *30*, 702–713. [CrossRef] [PubMed]
17. Sharma, M. Effect of Hydrocolloid Type on Rheological and Sensory Properties of Pureed Carrots. Master's Thesis, University of Guelph, Guelph, ON, Canada, 2015.
18. Sharma, M.; Kristo, E.; Corredig, M.; Duizer, L. Effect of hydrocolloid type on texture of pureed carrots: Rheological and sensory measures. *Food Hydrocoll.* **2017**, *63*, 478–487. [CrossRef]
19. Schlich, P. Temporal Dominance of Sensations (TDS): A new deal for temporal sensory analysis. *Curr. Opin. Food Sci.* **2017**, *15*, 38–42. [CrossRef]
20. Varela, P.; Antúnez, L.; Carlehög, M.; Alcaire, F.; Castura, J.C.; Berget, I.; Giménez, A.; Næs, T.; Ares, G. What is dominance? An exploration of the concept in TDS tests with trained assessors and consumers. *Food Qual. Preference* **2018**, *64*, 72–81. [CrossRef]
21. Pineau, N.; de Bouillé, A.G.; Lepage, M.; Lenfant, F.; Schlich, P.; Martin, N.; Rytz, A. Temporal Dominance of Sensations: What is a good attribute list? *Food Qual. Preference* **2012**, *26*, 159–165. [CrossRef]
22. Pineau, N.; Cordelle, S.; Schlich, P. Temporal Dominance of Sensations: A new technique to record several sensory attributes simultaneously over time. In Proceedings of the 5th Pangborn Sensory Science Symposium, Boston, MA, USA, 20–24 July 2003; p. 121.
23. Miele, N.A.; Di Monaco, R.; Dell'Amura, F.; Rega, M.F.; Picone, D.; Cavella, S. A preliminary study on the application of natural sweet proteins in agar-based gels. *J. Texture Stud.* **2017**, *48*, 103–113. [CrossRef]
24. Di Monaco, R.; Galiñanes Plaza, A.; Miele, N.A.; Picone, D.; Cavella, S. Temporal sweetness profile of MNEI protein in gelled model systems. *J. Sens. Stud.* **2016**, *31*, 382–392. [CrossRef]
25. Devezeaux de Lavergne, M.; Tournier, C.; Bertrand, D.; Salles, C.; van de Velde, F.; Stieger, M. Dynamic texture perception, oral processing behaviour and bolus properties of emulsion-filled gels with and without contrasting mechanical properties. *Food Hydrocoll.* **2016**, *52*, 648–660. [CrossRef]
26. Varela, P.; Pintor, A.; Fiszman, S. How hydrocolloids affect the temporal oral perception of ice cream. *Food Hydrocoll.* **2014**, *36*, 220–228. [CrossRef]
27. Castura, J.C.; Antúnez, L.; Giménez, A.; Ares, G. Temporal Check-All-That-Apply (TCATA): A novel dynamic method for characterizing products. *Food Qual. Preference* **2016**, *47*, 79–90. [CrossRef]
28. Reyes, M.M.; Castura, J.C.; Hayes, J.E. Characterizing dynamic sensory properties of nutritive and nonnutritive sweeteners with temporal check-all-that-apply. *J. Sens. Stud.* **2017**, *32*, e12270. [CrossRef]
29. Nguyen, Q.C.; Næs, T.; Varela, P. When the choice of the temporal method does make a difference: TCATA, TDS and TDS by modality for characterizing semi-solid foods. *Food Qual. Prefer.* **2018**, *66*, 95–106. [CrossRef]
30. Esmerino, E.A.; Castura, J.C.; Ferraz, J.P.; Tavares Filho, E.R.; Silva, R.; Cruz, A.G.; Freitas, M.Q.; Bolini, H.M.A. Dynamic profiling of different ready-to-drink fermented dairy products: A comparative study using Temporal Check-All-That-Apply (TCATA), Temporal Dominance of Sensations (TDS) and Progressive Profile (PP). *Food Res. Int.* **2017**, *101*, 249–258. [CrossRef]
31. Ares, G.; Jaeger, S.R.; Antúnez, L.; Vidal, L.; Giménez, A.; Coste, B.; Picallo, A.; Castura, J.C. Comparison of TCATA and TDS for dynamic sensory characterization of food products. *Food Res. Int.* **2015**, *78*, 148–158. [CrossRef]
32. Richman, J.W. *Pureed Foods with Substance and Style*; Aspen Publishers, Inc.: Gaithersburg, MD, USA, 1994; ISBN 0-8342-0554-8.

33. Dietary Guidelines for Americans, Older Adults Health Facts. Available online: https://health.gov/dietaryguidelines/dga2005/toolkit/default.htm#older_adults (accessed on 15 May 2019).
34. Tang, J.; Larsen, D.S.; Ferguson, L.; James, B.J. Textural Complexity Model Foods Assessed with Instrumental and Sensory Measurements. *J. Texture Stud.* **2017**, *48*, 9–22. [CrossRef]
35. Pineau, N.; Schlich, P.; Cordelle, S.; Mathonnière, C.; Issanchou, S.; Imbert, A.; Rogeaux, M.; Etiévant, P.; Köster, E. Temporal Dominance of Sensations: Construction of the TDS curves and comparison with time—Intensity. *Food Qual. Preference* **2009**, *20*, 450–455. [CrossRef]
36. Peltier, C.; Visalli, M.; Schlich, P. Canonical Variate Analysis of Sensory Profiling Data. *J. Sens. Stud.* **2015**, *30*, 316–328. [CrossRef]
37. Lenfant, F.; Loret, C.; Pineau, N.; Hartmann, C.; Martin, N. Perception of oral food breakdown. The concept of sensory trajectory. *Appetite* **2009**, *52*, 659–667. [CrossRef] [PubMed]
38. Castura, J.C. tempR: Temporal sensory data analysis. In Proceedings of the AgroStat (14th Symposium on Statistical Methods for the Food Industry), Lausanne, Switzerland, 21–24 March 2016.
39. Fernández, C.; Canet, W.; Alvarez, M.D. Quality of mashed potatoes: Effect of adding blends of kappa-carrageenan and xanthan gum. *Eur. Food Res. Technol.* **2009**, *229*, 205–222. [CrossRef]
40. Terpstra, M.E.J. Oral Texture Perception of Semisolid Foods in Relation to Physicochemical Properties. Ph.D. Thesis, Wageningen University, Wageningen, The Netherlands, 2008.
41. Choi, H.; Mitchell, J.R.; Gaddipati, S.R.; Hill, S.E.; Wolf, B. Shear rheology and filament stretching behaviour of xanthan gum and carboxymethyl cellulose solution in presence of saliva. *Food Hydrocoll.* **2014**, *40*, 71–75. [CrossRef]
42. BeMiller, J.N. Pasting, paste, and gel properties of starch–hydrocolloid combinations. *Carbohydr. Polym.* **2011**, *86*, 386–423. [CrossRef]
43. Krystyjan, M.; Sikora, M.; Adamczyk, G.; Tomasik, P. Caramel sauces thickened with combinations of potato starch and xanthan gum. *J. Food Eng.* **2012**, *112*, 22–28. [CrossRef]
44. Devezeaux de Lavergne, M.; van Delft, M.; van de Velde, F.; van Boekel, M.A.J.S.; Stieger, M. Dynamic texture perception and oral processing of semi-solid food gels: Part 1: Comparison between QDA, progressive profiling and TDS. *Food Hydrocoll.* **2015**, *43*, 207–217. [CrossRef]
45. Saint-Eve, A.; Panouillé, M.; Capitaine, C.; Déléris, I.; Souchon, I. Dynamic aspects of texture perception during cheese consumption and relationship with bolus properties. *Food Hydrocoll.* **2015**, *46*, 144–152. [CrossRef]

Article

Dynamic Oral Texture Properties of Selected Indigenous Complementary Porridges Used in African Communities

James Makame [1,*], Tanita Cronje [2], Naushad M. Emmambux [1] and Henriette De Kock [1]

1 Department of Consumer and Food Sciences, University of Pretoria, Private Bag X20, Hatfield, Pretoria 0028, South Africa; naushad.emmambux@up.ac.za (N.M.E.); riette.dekock@up.ac.za (H.D.K.)

2 Department of Statistics, University of Pretoria, Private Bag X20, Hatfield, Pretoria 0028, South Africa; tanita.cronje@up.ac.za

* Correspondence: makamejames@gmail.com

Received: 1 May 2019; Accepted: 13 June 2019; Published: 21 June 2019

Abstract: Child malnutrition remains a major public health problem in low-income African communities, caused by factors including the low nutritional value of indigenous/local complementary porridges (CP) fed to infants and young children. Most African children subsist on locally available starchy foods, whose oral texture is not well-characterized in relation to their sensorimotor readiness. The sensory quality of CP affects oral processing (OP) abilities in infants and young children. Unsuitable oral texture limits nutrient intake, leading to protein-energy malnutrition. The perception of the oral texture of selected African CPs (n = 13, Maize, Sorghum, Cassava, Orange-fleshed sweet potato (OFSP), Cowpea, and Bambara) was investigated by a trained temporal-check-all-that-apply (TCATA) panel (n = 10), alongside selected commercial porridges (n = 19). A simulated OP method (Up-Down mouth movements- munching) and a control method (lateral mouth movements- normal adult-like chewing) were used. TCATA results showed that Maize, Cassava, and Sorghum porridges were initially too thick, sticky, slimy, and pasty, and also at the end not easy to swallow even at low solids content—especially by the Up-Down method. These attributes make CPs difficult to ingest for infants given their limited OP abilities, thus, leading to limited nutrient intake, and this can contribute to malnutrition. Methods to improve the texture properties of indigenous CPs are needed to optimize infant nutrient intake.

Keywords: oral processing; TCATA; texture; malnutrition; sensorimotor readiness; complementary porridge; infant

---

1. Introduction

Food oral processing (OP), the manipulation and break down of food inside the mouth up to the moment of swallowing [1,2], plays a key and important role in sensory perception, consumer acceptance, and food intake [3]. It is a dynamic process; however, the scientific explanations regarding the sensory quality of baby food is lacking [4], particularly the effects of interactions between complementary porridge (CP)'s texture properties and the oral physiology of infants. Oral texture perception remains poorly understood, despite it being a key driver of food acceptance or rejection [5]. Most infants in Africa are nourished on low-cost complementary porridges (CPs) prepared from starchy plant materials (cereals, roots, tubers, and legumes) [6,7]. Such porridges are often thick even at low (about 8–10%) solids content [8–10]. In infants and young children, physiological capacity (chewing, salivation, and digestion), sensory quality and oral motor skills are important determinants of food choice [11]. Food texture is a perception arising from the interactions of the food physical structure with mechanoreceptors in the oral cavity [5]. Inappropriate porridge viscosity may compromise nutrient intake and lead to child malnutrition, a major public health problem, especially in low- to middle-income countries [12].

Children of different ages have different OP abilities to successfully chew and swallow foods of different physical forms [13]. The process of bolus formation is under the coordinated action of mastication (reduction of food in particles), salivation (lubrication of particles), and tongue movements (agglomeration of particles with saliva and swallowing) [1] and depends, thus, mainly on the development of the infant masticatory apparatus. For ingestion and break down of solid foods, children need to acquire specific feeding skills which require more effort than the oral manipulation of liquid milk [14]. The acceptance of food with a given texture (in this context defined as the infant's ability to swallow the food [11]) strongly depends on the acquisition of feeding skills, which can develop differently in children of the same age [15]. At 12 months, munching/chewing behavior is well-established and continues to develop and optimize by 2–3 years [16]. However, the age of chewing maturation (i.e., transition of up and down movements of the jaw to rotary movements) remains not so clear and is estimated to be later than 3 years [14]. Good quality CPs must have low viscosity, high nutrient density, appropriate texture, and a consistency that allows for easy consumption by infants and young children [17,18]. At the beginning of complementary feeding (4–6 months), infants prefer soft and smooth textured foods as they require limited oral manipulation before being swallowed [11]. Commercial infant porridges are considered to have the optimal quality, but they are too expensive for many poor families [19].

Bolus flow and ease of swallowing depend on the rheology of the bolus and the oral physiological conditions of the consumer [20,21]. Infants have limited dentition, weak masticatory muscles, and reduced tongue or pharyngeal muscle strength [22]. For safe and comfortable consumption, the porridge consistency should be matched with the child's oromotor readiness [23]. At present, there is paucity of research on the relationship between the in-mouth perceived texture of indigenous CPs and the sensorimotor development (oromotor readiness) of infants (6–12 months) and young children (13–24 months). Yet, infants and toddlers present a challenge to sensory and consumer researchers because of their inability to communicate verbally, limited cognitive abilities, and very low attention span [24,25]. Sensory testing with infants and young children, therefore, has often employed indirect approaches. As an example, descriptive sensory profiling has been used to evaluate the sensory quality of baby foods (purees) [4].

For preference evaluation, parents' liking is important in deciding if a given CP would be suitable for their infants [26]. The primary caretaker (typically the mother) interprets the behavior of the child during food tasting and rates acceptance on a hedonic scale [11,27,28]. The adult also tastes and rate the samples after the child, providing a control and confirmation of the acceptability of the samples [25]. In most studies, mothers are often asked to report on the presence/absence of positive and negative behaviors and on infant's food liking during feeding. Alternative testing approaches employed include parents completing an Infant Behavior Questionnaire and rating of videotapes of infants' facial reactions to foods rated [29].

Data from rheological studies (not included in this paper) have shown that indigenous porridge samples at very low solids content (Maize 8.1%, Sorghum 8.4%, Cassava 6.4%, Bambara groundnut 10.7%, and Cowpea 10.1%) exceeded the recommended CP viscosity limit (3 Pa·s at 40 °C and shear rate of 50 s^{-1}, [30,31]) for infants and children below 3 years of age. When starch is heated in water, it swells, gelatinizes, and pastes to form a thick gruel [8–10]. Infants and young children have difficulty to consume and swallow a viscous porridge due to their limited oromotor capacity [32]. The thickness or viscosity of shear-thinning foods is perceived by mechanoreceptors in the mouth, and oral thickness perception depends on the in-mouth shear stress applied and the resultant shear rate [33]. At the critical solids concentration (c*), the gelatinized, amorphous random starch polymer coils come in contact with one another, eventually overlapping at the entanglement concentration (c_e) [34]. Weight for weight, polymers with larger molecules display more effective molecular entanglements [35,36], and are generally perceived as more viscous and thick compared to those with smaller moleules. High viscosity in CP elicits high lingual swallowing pressure [37]. Thicker and harder foods are eaten at a slower rate, often requiring smaller bite sizes and more chewing time in the mouth before swallowing compared to softer foods [38]. The formation of a bolus that can be safely swallowed is a complex oral process [11], and infants and young children have limited oral capacity to perform this process.

The aim of this study was to characterize the texture of selected indigenous CPs typically used for feeding infants and young children aged 6–24 months in African communities during OP (therefore dynamic), in order to make recommendations for optimizing their oral texture to improve nutrient intake. A trained sensory panel consisting of adults was used because infants are not capable of carrying out evaluation instructions and tasks expected in descriptive sensory evaluation. To understand the temporal in-mouth textural nuances, the design applied two different OP methods: a novel procedure (the Up-Down mouth movements- munching) that mimics how infants with limited OP abilities process food, and a control method (chewing with lateral mouth movements) representing normal adult OP.

2. Materials and Methods

2.1. Samples and Sample Preparation

Table 1 shows the descriptions of indigenous and commercial CP samples used in the study. To make flour, Bambara groundnut (*Vigna subterranean* (L.) Verdc) and Cowpea *(Vigna unguiculata)* seeds were first decorticated using a Tangential Abrasive Dehulling Device (TADD, Venables Machine Works, Saskatoon, Canada) and milled to <250 μm particle size flour (Laboratory mill 3100, Perten Instruments, Hägersten, Sweden). All indigenous CP samples were prepared as described by [39] with adaptations. A specific amount of flour was measured into an aluminum pot following Table 1, and 250 mL of cold water was added while stirring to form a uniform slurry. The flour quantities for each treatment represent the solids % determined from rheological experiments, that give the recommended porridge viscosity limit of 3 Pa·s (at 40 °C and shear rate of 50 s^{-1}). The pot was placed over a hot plate, and the remaining quantity of boiling water was slowly added to the slurry while continuously stirring. Once the mixture began to boil, the timer was started, and the porridge was cooked for 7 min with continuous stirring. Porridge samples were transferred to appropriately labeled containers placed over a water-bath maintained at 55 °C until serving. Commercial porridges were prepared, according to the manufacturers' instructions, by mixing with water or milk depending on the product. The processing techniques for commercial porridges allow them to remain thin at a higher solids % compared to indigenous porridges.

Table 1. Description of complementary porridges (CPs) evaluated for oral texture by the trained TCATA sensory panel.

Porridge Indigenous/Local	Flour (g)	Water (g)	Solids (%) #	Description and Source
Maize	40 80 100	960 920.0 900	4 8 10	Super maize meal (commercially processed) from the local supermarket (Pretoria, RSA)
Sorghum	40 80 100	960 920 900	4 8 10	Super mabela flour (commercially processed) from local supermarket (Pretoria, RSA)
Bambara	100	900	10	Dry Seeds, cream cultivar, Mbare Produce market (Harare, Zimbabwe)
Cowpea	100	900	10	Commercial seeds, Agrinawa cultivar Agricol (Pty) Ltd. (Pretoria, RSA)
Cassava	40 60 100	960 940 900	4 6 10	High-quality cassava (84.4% starch), Thai Farm International (Ogun, Nigeria)
OFSP (Orange-fleshed sweet potato)	100 160	900 840	10 16	Dried with electric dryer (60 °C, 6–8 h), Exilite 499 cc (Tzaneen, Limpopo, RSA)
Commercial Porridges (Code)	**Age (Months)**	**Flour:Liquid (g:mL)**	**Solids (%)**	**Description/Manufacturers Instructions Guide ****
A1-Reference	6 to 24	50:150	25.0	Enzyme-hydrolyzed cereal (maize 62%), add water
A2	6 to 24	50:150	25.0	Enzyme-hydrolyzed cereal (rice 63%), add water
A3	6 to 24	50:150	25.0	Enzyme-hydrolyzed cereal (wheat 61%), add water
C2	6 to 24	50:140	26.3	Oat flakes 32%, add water
F1	6 to 8 9 to 12 13 to 36	45:150 67:200 80:250	23.1 25.1 24.2	Enzyme-hydrolyzed cereal (wheat 51%), add water
F2	6 to 8 9 to 12 13 to 36	35:150 60:200 75:250	18.9 23.1 23.1	Enzyme-hydrolyzed cereal (rice 51%), add water

Table 1. *Cont.*

Porridge Indigenous/Local	Flour (g)	Water (g)	Solids (%) #	Description and Source
F3	9 to 12 13 to 36	67:200 80:250	25.1 24.2	Enzyme-hydrolyzed cereal (wheat, rice, corn, rye, barley 54%), add water
F4	9 to 12 13 to 36	67:200 80:250	25.1 24.2	Enzyme-hydrolyzed cereal (wheat, rice, corn, rye, barley 43%), add water
B1	6 to 24	50:160	35.1	Enzyme-hydrolyzed cereal (maize), add water
B2	6 to 24	50:160	35.1	Enzyme-hydrolyzed cereal (wheat), add water
C1	6 to 24	20:170	24.8	Whole oat flour 70%, banana flakes 30%, add milk
D	6 to 36	20:140	26.3	Maize flour minimum 86%, add milk
E1	6 to 12 13 to 36	25:200 35:280	26.7 25.8	Maize meal flour, 3 min cook with milk *** Cooking loss of 5%
E2	6 to 12 13 to 36	25:125 35:190	30.0 29.1	Sorghum flour (minimum 89%), add milk
G	13 to 36	20:80	32.8	Wheat flour, maize flour, soy flour, add milk

* NAN Optipro milk (Formula 2 for 6–12 and Formula 3 for 13–24 months, respectively) was prepared as per manufacturer (mixing 32 g milk powder with 200 mL pre-boiled luke-warm water) to give a 16% solids content milk. The "add-water" commercial samples contain whole or skim milk (23–40%). # determined from rheological experiments, such that the flour % in water gives the recommended cooked porridge viscosity ≤3 Pa·s (at 40 °C and shear rate of 50 s^{-1}). ** Information given on the product pack. *** Native maize meal flour with unhydrolyzed or non-depolymerized starch molecules.

2.2. Sensory Analysis

The sensory analysis of the porridges was conducted in individual booths under white light and standardized conditions at the University of Pretoria Sensory Evaluation Laboratory. The use of human subjects in the study was approved by the Faculty of Natural and Agricultural Sciences Ethics Review Committee at the University of Pretoria (EC 180000086). Each participant signed a consent form prior to taking part in the study.

Ten assessors (3 males and 7 females, aged 22–27 years) were selected and trained on the Temporal Check-All-That-Apply (TCATA) evaluation method in three sessions of 2 h each, according to the guidelines of the International Organization for Standardization (ISO) standard 8586:2012 [40]. Prior to training, panelists were screened for interest, availability, general health status, and product discrimination abilities. With TCATA, panelists taste the samples and select all the sensory attributes they perceived at each moment of the evaluation [41]. They are allowed to check several attributes, which enables them to describe sensory characteristics that are simultaneously perceived [42]. During training, the assessors familiarized themselves with the oral texture of the CPs, discussed, and agreed on 14 attributes (Table 2). The panel was also trained on the evaluation protocol and use of the data acquisition software, Compusense Cloud version 7.8.2 (Compusense Inc., Guelph, ON, Canada).

TCATA has often been used with a large number of consumers in evaluating different products [43,44]. However, coupled with training, fewer panelists (10–15) have also previously been used for the temporal profiling of products based on 8–10 attributes [41,45]. The current study used 10 panelists well trained in carrying out the TCATA task and a list of 14 attributes. Temporal methods are more cognitively demanding and usually rely on shorter lists [46,47].

Table 2. Definitions of the in-mouth texture attributes used during the evaluation of complementary porridges (CPs) by a trained Temporal Check-All-That-Apply (TCATA) sensory panel (n = 10).

No.	TCATA Attribute	Definition
1	Soft	Selected when little force is required to orally process and move around the mouth.
2	Smooth	Selected when the sample is perceived as smooth when squeezed between the palate and tongue [48].
3	Creamy	Selected when the sample is perceived as creamy, with a silky smooth sensation in the mouth [48].
4	Grainy	Selected when grainy particles are perceived in the mouth.
5	Too thick/Semi-solid	Viscosity perception of cooked maize meal pastes 15–20% solids in water. Similar to mashed potato [49].
6	Thick	Viscosity perception of cooked maize meal paste (10–15% solids in water). Selected when the sample is perceived as thick (viscous) as opposed to thin like a fluid [48].
7	Thin	Selected when the sample is perceived as thin andfluid-likeas opposed to thick (viscous).
8	Chewy	Selected when the sample requires a substantial number of chews before it is ready to swallow [50].
9	Sticky	Selected when the sample sticks to the teeth and palate [48].
10	Watery	Selected when the sample was perceived as thin and watery [49].
11	Easy to swallow	Selected when the sample requires little effort (exertion/force) to swallow [51].
12	Difficult to swallow	Selected when the sample requires a lot of effort (exertion/force) to swallow.
13	Slimy	Selected when the sample is perceived as slimy and slippery, a mildly sticky perception on the palate/tongue.
14	Pasty	Selected when the sample has the consistency of a (starch) paste, semi-solid with some stickiness.

The terms "watery" and "too thick" were used to anchor the two extremes of the sensory space for porridge viscosity informed by panel feedback.

Thirty-two CPs (Table 1) were evaluated in duplicate, using two different OP methods, over 7 days with a maximum of ten CPs per day. The CPs were first evaluated using the Up-Down method that mimics feeding in infants and young children with limited OP ability. Assessors moved their mouths only up and down, avoiding sideways movements and ensuring limited tongue movement.

Babies initially use immature feeding skills characterized by the up and down movements of the jaws, eventually transitioning to mature feeding skills defined by rotary jaw movements, which facilitate efficient chewing [14,23]. In the second method (experienced adult chewing called Normal), panelists orally processed the food in a normal adult way involving the lateral mouth and tongue movements, applying oral shear and chewing where necessary. The Up-Down method was always used first before the Normal method during the evaluation sessions because the former required more conscious procedural effort due to its artificial nature compared to normal adult oral processing.

All attributes were presented in a three-column format on the computer screen. The order position of attributes on the TCATA list was randomized across assessors, but the list order remained consistent for a given assessor across all samples [52]. Porridge samples (±20 g) were presented to assessors monadically at 40 °C in glass ramekins covered with aluminum foil following a random balanced order. The evaluation instructions requested the panelists to select all terms on the TCATA list (Table 2) that described the sensations experienced at a given time of evaluation (measurements per second). Concurrently, they had to deselect the terms that were no longer relevant to describe the sensation of a given sample at that moment during the evaluation. To begin the evaluation, assessors took a spoonful (±5 g) of porridge into their mouth, clicked "start" on the computer screen ($t = 0$ s), and proceeded according to the instructions. After 20 s, they were prompted to swallow the sample and continued to note the sensations until the end of the evaluation. The evaluation duration (30 s) was established through an iterative process with assessors during training. It was an estimate of the time required for assessors to orally process a spoonful of porridge from intake to swallow while eating like a baby (munching). A 1 min break was enforced between samples for assessors to rinse their palate with deionized water.

2.3. Data Analysis

Attribute citations proportions were calculated using a procedure described by [53], as the percentage of assessors who selected an attribute ('1') at any given moment (every 1 s) during the evaluation period. TCATA curves were plotted using statistical software R [54], package tempR (version 0.9.9.15.) [55], and the lines smoothed by the cubic smoothing spline function to reduce noise in the data. For each attribute, reference lines per treatment at every 1 s during the evaluation period were calculated according to [56], as the average across all other CPs excluding the one that this average is contrasted with. Principal Component Analysis (PCA)—a multivariate data analysis method for visualization of correlations between multiple quantitative observations and variables [57], was used to produce PCA product trajectories (biplot). Visualizing observations in a 2- or 3-dimensional space permits identification of uniform or atypical groups of observations. PCA can be considered as a projection method, which projects observations from a p-dimensional space with p variables to a k-dimensional space (where $k < p$) so as to conserve the maximum amount of information from the initial dimensions [57].

For the average citation proportions, TCATA data were divided into 3 equal time slices of 10 s each (Initial: 1–10 s; Middle: 11–20 s; End: 21–30 s), and the mean values were obtained for each attribute at each time slice, as the proportion of the 10 s evaluation time that the attribute was selected. For example, if an assessor selected chewy for a duration of 8 s and creamy for a duration of 2 s, then the citation proportions for chewy would be 8/10 = 0.8, and for creamy would be 2/10 = 0.2. Data were analyzed using a mixed effects ANOVA model:

$$\text{AUC} = \text{Porridge Sample} + \text{OP Method} + \text{Porridge Sample} \times \text{OP Method} + \text{Assessor}, \quad (1)$$

where AUC refers to the area under the citation proportion by time (s) curve (average citations proportions) for each attribute in each treatment. The Sample, Method, and their two-way interaction effects were the fixed factors, whereas Assessors was a random effect. Multiple comparisons were done using Fischer's Least Significant Difference (LSD) test to detect which sample pairs were significantly different. XLStat software (version 2019.1.2) (Addinsoft, Long Island, New York, NY, USA) [58] was used for the analysis. Citation proportions (range: 0 to 1) correspond to the AUC in Time Intensity studies [48]. In order to test if the porridge samples differed significantly with respect to presence or absence of the 14 specific oral texture attributes, the data, pooled over OP methods and replications, were submitted to Cochran's Test, as described by [45]. Cochran's Q is used for testing $k = 2$ or more matched sets, where a binary response (e.g., 0 or 1) is recorded from each category within each subject. Cochran's Q test tests the null hypothesis that the proportion of "successes" is the same in all groups versus the alternative hypothesis that the proportion is different in at least one of the groups. For pairwise differences between each treatment within a time-segment, Marascuilo's Test was used [59]. The Cochran's Q test followed by Marascuilo's test was done in XLSTAT software (version 2019.1.2) (Addinsoft, Long Island, New York, NY, USA) [58].

3. Results

3.1. Dynamic Oral Texture: TCATA Product Profiles

Figure 1A–F shows TCATA curves for the three most common types of African CP (Maize, Sorghum, and Cassava, all 10% solids). When evaluated by the Normal OP method (Figure 1A,C,E), these porridges were perceived ($p < 0.05$) as more thick or too thick, more sticky, and less creamy during the initial (1–10 s) and middle (11–20 s) phases in comparison with the mean (dotted lines) of all other samples. In the Up-Down method, maize porridge (Figure 1B) was thicker ($p < 0.05$) for a much longer time period, with much higher citation proportions than in the Normal OP method (Figure 1C). The panel also perceived Cassava porridge as thick, too thick, sticky, slimy, and not creamy by both OP methods (Figure 1E,F). In the Up-Down method (Figure 1F), however, the Cassava porridge attributes were more prevalent and more persistent, being further described as significantly not thin (between 10 s and 15 s) and less easy to swallow compared to the rest of the samples.

The Cochran's Q test was carried out on the TCATA data collected during the 30 s evaluation period for the indigenous CPs and a commercial reference (Table 1 sample A1). The frequency of perception (citation proportion) for attributes used differed in time ($p \leq 0.05$) as a function of Samples (Table 3). The 6th, 16th, and 26th s time moments were taken to represent three phases of OP during the evaluation. Initially and mid-way during OP, assessors described Maize and Sorghum porridges (8–10% solids) and Cassava 10% as thick, and even too thick, relative to the other CPs ($p \leq 0.05$). Cassava CPs (6 and 10% solids) were also characterized as significantly ($p < 0.05$) sticky and pasty and together with Bambara groundnut and Cowpea (all perceived as slimy $p \leq 0.05$) when compared with the other porridges. CP attribute differences declined as OP progressed, meaning that the CPs became more similar towards the end of OP. During swallowing, more panelists perceived Cassava 10% as thick, sticky, pasty, and slimy compared to the other porridges ($p \leq 0.05$). The slimy texture was more frequently perceived in all legume porridges (Bambara groundnut and Cowpea) but not in the cereal and orange-fleshed sweet potato (OFSP) based porridges.

A

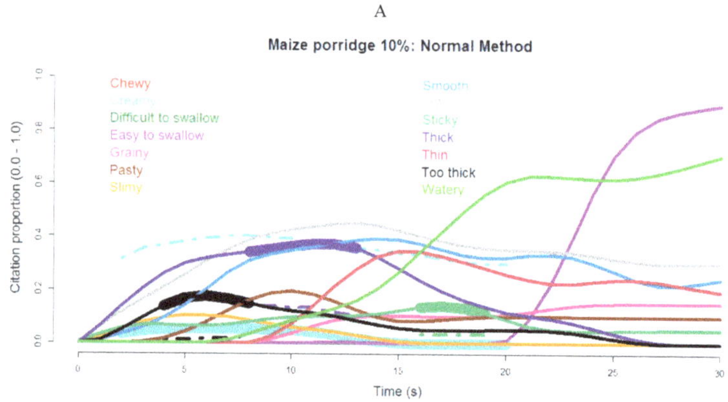

Figure 1. *Cont.*

B

Maize porridge 10%: UpDown Method

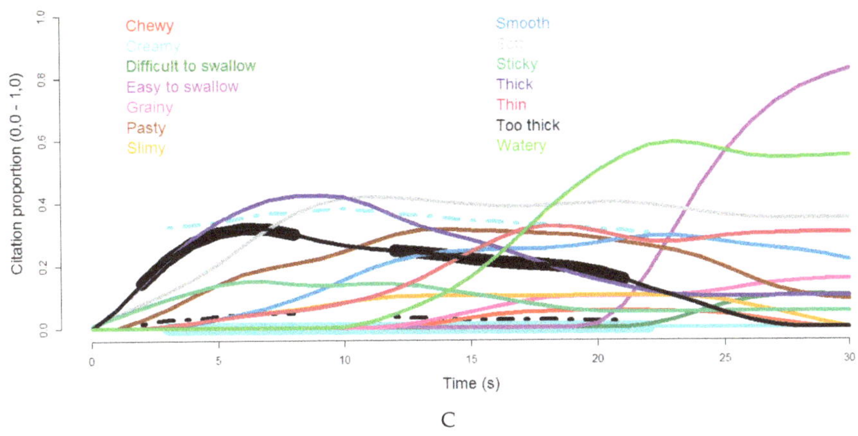

C

Sorghum porridge 10%: Normal Method

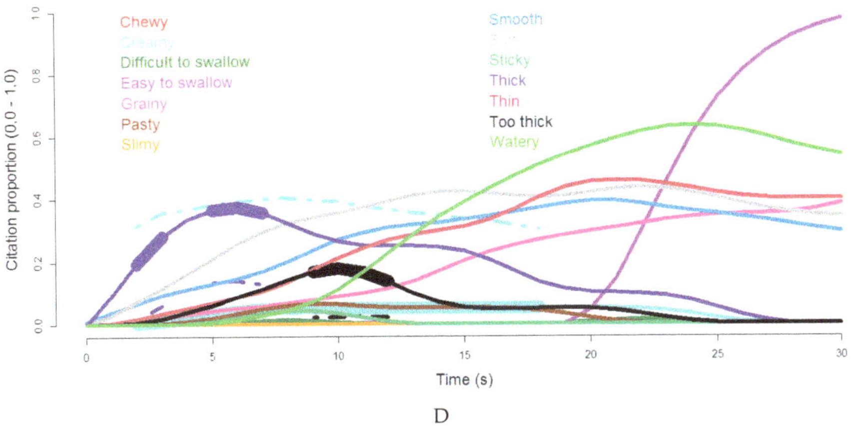

D

Sorghum porridge 10%: UpDown Method

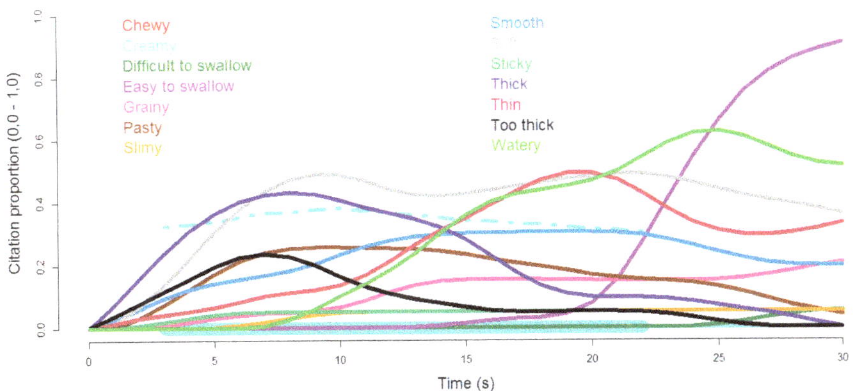

Figure 1. *Cont.*

E

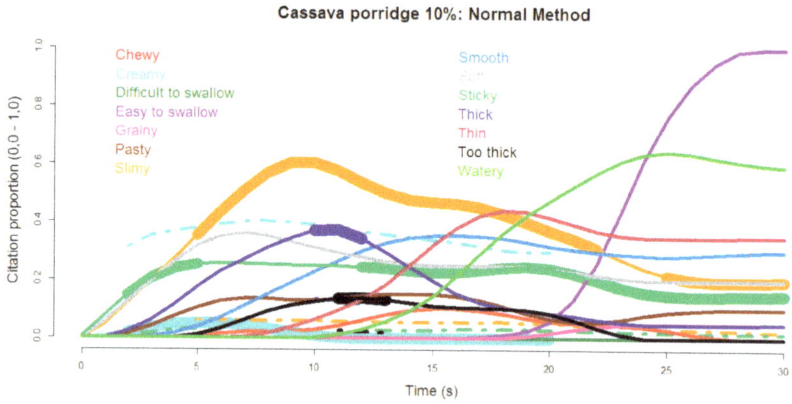

Cassava porridge 10%: Normal Method

F

Cassava porridge 10%: UpDown Method

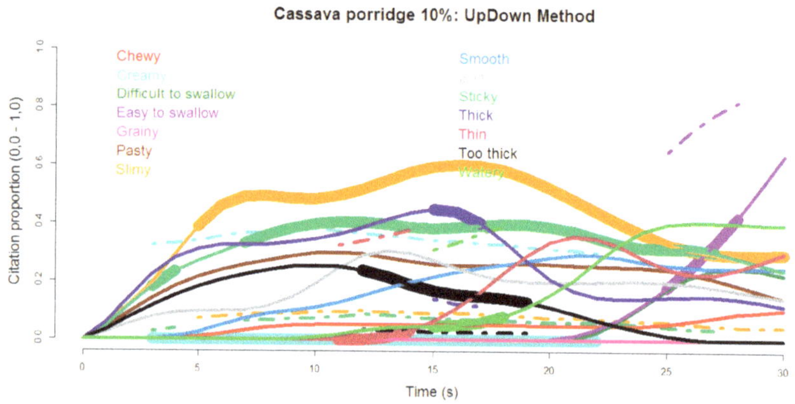

Figure 1. Temporal Check-All-That-Apply (TCATA) texture attribute curves for maize, sorghum, and cassava complimentary porridges (CPs) (10% solids). Attribute reference lines (represented as dotted lines in the figures) are shown only during periods of significant differences ($p \leq 0.05$) in citation proportion for that porridge compared to the mean of all other CPs. Significant reference line segments are contrasted with highlighted thicker sections of attribute curves for convenient visualization. The letter **A–F** represent maize, sorghum and cassava CPs evaluated by the Normal and the Up-Down OP method respectively.

When applying a 2-way mixed model ANOVA to evaluate the effect of Sample and OP method (Table 4), similar results were noted. Maize and Sorghum (8–10% solids) and Cassava (10% solids) were characterized by the panel as significantly thick or too thick, sticky, slimy, and pasty compared to the rest of the samples, in all three time-slices (only results for initial and end time-slices are shown in Table 4). The Up-Down OP method gave significantly higher thick and too thick citation proportions compared to the Normal OP method for Maize (8–10% solids), Sorghum (8–10% solids), and Cassava (10% solids) porridges. As in Cochran's Q test, slimy texture during swallowing was more perceived in cassava and leguminous CPs, with Cassava 10% described as not easy to swallow by a higher proportion of panelists.

Table 3. The effect of porridge type on the citation proportions for some oral texture attributes at three different moments during evaluation by a trained Temporal Check-All-That-Apply (TCATA) sensory panel (n = 10): Data analyzed using Cochran Q test followed by Marascuillo's pairwise comparison test.

Porridge Type	Beginning (6 s)					Middle (16 s)					End (26 s)			
	Thick	Sticky	Too thick	Pasty	Slimy	Thick	Sticky	Too thick	Pasty	Slimy	Thick	Sticky	Pasty	Slimy
Bambara 10%	0.05 ^{a}	0.03 ^{ab}	0.00 ^{a}	0.03 ^{ab}	0.25 ^{b}	0.00 ^{a}	0.00 ^{a}	0.00 ^{a}	0.05 ^{ab}	0.20 ^{b}	0.00 ^{a}	0.00 ^{a}	0.03 ^{ab}	0.15 ^{b}
Cowpea 10%	0.08 ^{a}	0.13 ^{cd}	0.00 ^{a}	0.05 ^{abc}	0.35 ^{bc}	0.05 ^{ab}	0.08 ^{ab}	0.03 ^{a}	0.08 ^{abc}	0.20 ^{b}	0.03 ^{ab}	0.05 ^{a}	0.05 ^{ab}	0.13 ^{b}
Cassava 4%	0.00 ^{a}	0.10 ^{bcd}	0.00 ^{a}	0.03 ^{ab}	0.40 ^{bc}	0.00 ^{a}	0.05 ^{ab}	0.00 ^{a}	0.00 ^{a}	0.25 ^{bc}	0.00 ^{a}	0.05 ^{a}	0.00 ^{a}	0.13 ^{b}
Cassava 6%	0.08 ^{a}	0.18 ^{de}	0.00 ^{a}	0.15 ^{cd}	0.43 ^{c}	0.05 ^{ab}	0.05 ^{ab}	0.00 ^{a}	0.05 ^{ab}	0.38 ^{c}	0.00 ^{a}	0.05 ^{a}	0.03 ^{ab}	0.18 ^{bc}
Cassava 10%	0.30 ^{b}	0.25 ^{e}	0.10 ^{b}	0.20 ^{d}	0.50 ^{c}	0.33 ^{f}	0.30 ^{c}	0.13 ^{b}	0.20 ^{c}	0.55 ^{d}	0.10 ^{c}	0.25 ^{b}	0.18 ^{cd}	0.25 ^{c}
Maize 4%	0.00 ^{a}	0.03 ^{ab}	0.00 ^{a}	0.00 ^{a}	0.05 ^{a}	0.00 ^{a}	000 ^{a}	0.00 ^{a}	0.00 ^{a}	003 ^{a}	0.00 ^{a}	0.00 ^{a}	0.00 ^{a}	0.00 ^{a}
Maize 8%	0.40 ^{b}	0.05 ^{abc}	0.05 ^{ab}	0.03 ^{ab}	0.08 ^{a}	0.15 ^{cd}	0.03 ^{a}	0.05 ^{a}	0.05 ^{ab}	0.00 ^{a}	0.03 ^{ab}	0.00 ^{a}	0.03 ^{ab}	0.00 ^{a}
Maize 10%	0.33 ^{b}	0.10 ^{bcd}	0.25 ^{d}	0.13 ^{bcd}	0.08 ^{a}	0.23 ^{de}	0.13 ^{b}	0.13 ^{b}	0.20 ^{c}	0.05 ^{a}	0.05 ^{b}	0.05 ^{a}	0.13 ^{bcd}	0.03 ^{a}
Sorghum 4%	0.05 ^{a}	0.00 ^{a}	0.00 ^{a}	0.00 ^{a}	0.05 ^{a}	0.00 ^{a}	0.00 ^{a}	0.00 ^{a}	005 ^{ab}	0.05 ^{a}	0.00 ^{a}	0.00 ^{a}	0.03 ^{ab}	0.00 ^{a}
Sorghum 8%	0.30 ^{b}	0.03 ^{ab}	0.05 ^{ab}	0.00 ^{a}	0.03 ^{a}	0.10 ^{bc}	0.08 ^{ab}	005 ^{a}	0.08 ^{abc}	0.03 ^{a}	0.05 ^{b}	0.03 ^{a}	0.08 ^{abc}	0.03 ^{a}
Sorghum 10%	0.40 ^{b}	0.03 ^{ab}	0.18 ^{c}	0.15 ^{cd}	0.00 ^{a}	0.25 ^{ef}	0.03 ^{a}	0.05 ^{a}	0.13 ^{abc}	0.03 ^{a}	0.03 ^{ab}	0.00 ^{a}	0.05 ^{ab}	0.03 ^{a}
OFSP 10%	0.00 ^{a}	0.00 ^{a}	0.00 ^{a}	0.08 ^{abc}	0.00 ^{a}	0.00 ^{a}	0.00 ^{a}	0.00 ^{a}	0.10 ^{abc}	0.03 ^{a}	0.03 ^{ab}	0.00 ^{a}	0.10 ^{abcd}	0.03 ^{a}
OFSP 16%	0.10 ^{a}	0.03 ^{ab}	0.00 ^{a}	0.05 ^{abc}	0.03 ^{a}	0.00 ^{a}	0.08 ^{ab}	0.00 ^{a}	0.15 ^{bc}	0.03 ^{a}	0.00 ^{a}	0.05 ^{a}	0.20 ^{d}	0.03 ^{a}
A1 *	0.00 ^{a}	0.00 ^{a}	0.00 ^{a}	0.03 ^{ab}	0.03 ^{a}	0.00 ^{a}	0.03 ^{a}	0.00 ^{a}	0.10 ^{abc}	0.05 ^{a}	0.00 ^{a}	0.00 ^{a}	0.10 ^{abcd}	0.03 ^{a}

^{abcdef} Citation proportions with different letters within a column represent significant differences among treatments at $p \leq 0.05$ as analyzed using Marascuillo's test. Responses were pooled across replicates, and oral processing method (40 TCATA runs × 14 samples) was used to determine the citation proportion values. A1 * is a commercial reference porridge.

Table 4. Effect of complimentary porridge (CP) type and oral processing (OP) method on citation proportions for texture attributes during the initial (1–10 s) and ending (21–30 s) phases of evaluation by the trained Temporal Check-All-That-Apply (TCATA) panel (n = 10). Main effects ANOVA followed by Fisher's Least Significant Difference (LSD) test for pairwise comparisons.

Porridge-Sample	Oral-Method	Initial: 1–10 s					End: 21–30 s				
		Thick	Too thick	Sticky	Slimy	Pasty	Thick	Too Thick	Slimy	Pasty	Swallow (+)
Maize 4%	Normal	0.00 ^{a}	0.00 ^{a}	0.02 ^{ab}	0.04 ^{a}	0.00 ^{a}	0.00 ^{a}	0.00 ^{a}	0.00 ^{a}	0.00 ^{a}	0.64 ^{bcdefghij}
	Up-Down	0.00 ^{a}	0.00 ^{a}	0.01 ^{a}	0.06 ^{abc}	0.00 ^{a}	0.00 ^{a}	0.00 ^{a}	0.01 ^{a}	0.00 ^{a}	0.77 ^{ij}
Maize 8%	Normal	0.23 ^{def}	0.03 ^{ab}	0.05 ^{abc}	0.02 ^{a}	0.00 ^{a}	0.05 ^{abc}	0.01 ^{ab}	0.00 ^{a}	0.00 ^{a}	0.76 ^{hij}
	Up-Down	0.37 ^{g}	0.03 ^{ab}	0.07 ^{abcd}	0.05 ^{ab}	0.05 ^{abc}	0.01 ^{bc}	0.03 ^{ab}	0.00 ^{a}	0.05 ^{abc}	0.59 ^{bcdefgh}
Maize 10%	Normal	0.26 ^{defg}	0.11 ^{cd}	0.07 ^{abcd}	0.07 ^{abc}	0.08 ^{abcd}	0.04 ^{abc}	0.02 ^{ab}	0.00 ^{a}	0.10 ^{abcd}	0.62 ^{bcdefghi}
	Up-Down	0.31 ^{efg}	0.20 ^{e}	0.10 ^{bcd}	0.00 ^{a}	0.13 ^{bcde}	0.01 ^{cd}	0.00 ^{a}	0.10 ^{abcd}	0.18 ^{cde}	0.53 ^{bcde}
Sorghum 4%	Normal	0.01 ^{a}	0.00 ^{a}	0.00 ^{a}	0.02 ^{a}	0.01 ^{a}	0.00 ^{a}	0.00 ^{a}	0.02 ^{ab}	0.05 ^{abc}	0.65 ^{cdefghij}
	Up-Down	0.02 ^{a}	0.00 ^{a}	0.00 ^{a}	0.04 ^{a}	0.01 ^{a}	0.00 ^{a}	0.00 ^{a}	0.01 ^{a}	0.00 ^{a}	0.63 ^{bcdefghij}
Sorghum 8%	Normal	0.20 ^{cde}	0.04 ^{abc}	0.04 ^{abc}	0.00 ^{a}	0.02 ^{a}	0.00 ^{a}	0.01 ^{a}	0.00 ^{a}	0.07 ^{abc}	0.56 ^{bcdefg}
	Up-Down	0.36 ^{g}	0.04 ^{abc}	0.04 ^{abc}	0.04 ^{a}	0.05 ^{ab}	0.09 ^{bcd}	0.03 ^{ab}	0.05 ^{abcd}	0.05 ^{abc}	0.59 ^{bcdefgh}
Sorghum 10%	Normal	0.28 ^{defg}	0.08 ^{bc}	0.02 ^{ab}	0.00 ^{a}	0.03 ^{a}	0.04 ^{abc}	0.01 ^{a}	0.00 ^{a}	0.00 ^{a}	0.67 ^{defghij}
	Up-Down	0.32 ^{fg}	0.17 ^{de}	0.03 ^{abc}	0.01 ^{a}	0.16 ^{de}	0.06 ^{abc}	0.02 ^{ab}	0.05 ^{abcd}	0.11 ^{abcd}	0.63 ^{bcdefghij}
Bambara 10%	Normal	0.02 ^{a}	0.00 ^{a}	0.01 ^{a}	0.18 ^{cd}	0.00 ^{a}	0.00 ^{a}	0.00 ^{a}	0.15 ^{bcde}	0.00 ^{a}	0.75 ^{hij}
	Up-Down	0.03 ^{a}	0.00 ^{a}	0.01 ^{a}	0.18 ^{cd}	0.05 ^{ab}	0.00 ^{a}	0.00 ^{a}	0.15 ^{bcde}	0.05 ^{abc}	0.76 ^{hij}
Cowpea 10%	Normal	0.07 ^{ab}	0.00 ^{a}	0.09 ^{abcd}	0.24 ^{de}	0.01 ^{a}	0.00 ^{a}	0.00 ^{a}	0.11 ^{abcde}	0.05 ^{abc}	0.70 ^{efghij}
	Up-Down	0.05 ^{a}	0.00 ^{a}	0.09 ^{abcd}	0.28 ^{def}	0.06 ^{abc}	0.07 ^{abcd}	0.00 ^{a}	0.16 ^{cde}	0.08 ^{abc}	0.52 ^{bcd}
Cassava 4%	Normal	0.00 ^{a}	0.00 ^{a}	0.08 ^{abcd}	0.28 ^{def}	0.02 ^{a}	0.00 ^{a}	0.00 ^{a}	0.13 ^{abcde}	0.00 ^{a}	0.73 ^{ghij}
	Up-Down	0.00 ^{a}	0.00 ^{a}	0.08 ^{abcd}	0.35 ^{ef}	0.05 ^{ab}	0.00 ^{a}	0.00 ^{a}	0.18 ^{de}	0.00 ^{a}	0.55 ^{bcdef}
Cassava 6 %	Normal	0.05 ^{a}	0.00 ^{a}	0.12 ^{cde}	0.33 ^{ef}	0.07 ^{abcd}	0.00 ^{a}	0.00 ^{a}	0.19 ^{de}	0.00 ^{a}	0.67 ^{cdefhij}
	Up-Down	0.06 ^{a}	0.00 ^{a}	0.16 ^{def}	0.39 ^{f}	0.14 ^{cde}	0.01 ^{ab}	0.00 ^{a}	0.24 ^{de}	0.08 ^{abc}	0.60 ^{bcdefhi}
Cassava 10%	Normal	0.19 ^{bcd}	0.04 ^{abc}	0.22 ^{ef}	0.38 ^{f}	0.08 ^{abcd}	0.05 ^{abc}	0.01 ^{a}	0.24 ^{ef}	0.09 ^{abc}	0.70 ^{efghij}
	Up-Down	0.27 ^{defg}	0.17 ^{de}	0.26 ^{f}	0.35 ^{ef}	0.20 ^{e}	0.15 ^{d}	0.03 ^{ab}	0.35 ^{f}	0.22 ^{de}	0.26 ^{a}
OFSP 10%	Normal	0.00 ^{a}	0.00 ^{a}	0.00 ^{a}	0.00 ^{a}	0.06 ^{abc}	0.00 ^{a}	0.00 ^{a}	0.03 ^{abc}	0.13 ^{bcde}	0.56 ^{bcdefg}
	Up-Down	0.00 ^{a}	0.00 ^{a}	0.00 ^{a}	0.00 ^{a}	0.08 ^{abcd}	0.05 ^{abc}	0.00 ^{a}	0.00 ^{a}	0.05 ^{abc}	0.65 ^{cdefghij}
OFSP 16%	Normal	0.03 ^{a}	0.00 ^{a}	0.02 ^{ab}	0.00 ^{a}	0.03 ^{a}	0.00 ^{a}	0.00 ^{a}	0.00 ^{a}	0.13 ^{abcd}	0.47 ^{b}
	Up-Down	0.10 ^{abc}	0.00 ^{a}	0.05 ^{abc}	0.03 ^{a}	0.08 ^{abcd}	0.00 ^{a}	0.00 ^{a}	0.05 ^{abcd}	0.25 ^{e}	0.50 ^{bc}
A1	Normal	0.00 ^{a}	0.00 ^{a}	0.00 ^{a}	0.02 ^{a}	0.01 ^{a}	0.00 ^{a}	0.05 ^{b}	0.00 ^{a}	0.05 ^{abc}	0.68 ^{defghi}
	Up-Down	0.00 ^{a}	0.00 ^{a}	0.00 ^{a}	0.04 ^{a}	0.05 ^{abc}	0.00 ^{a}	0.00 ^{a}	0.05 ^{abcd}	0.15 ^{cde}	0.72 ^{fghi}
A2	Normal	0.08 ^{abc}	0.00 ^{a}	0.00 ^{a}	0.00 ^{a}	0.00 ^{a}	0.00 ^{a}	0.00 ^{a}	0.04 ^{abc}	0.05 ^{abc}	0.62 ^{bcdefghi}
	Up-Down	0.06 ^{a}	0.00 ^{a}	0.00 ^{a}	0.00 ^{a}	0.04 ^{ab}	0.05 ^{abc}	0.00 ^{a}	0.01 ^{a}	0.11 ^{abcd}	0.53 ^{bcde}
A3	Normal	0.00 ^{a}	0.00 ^{a}	0.00 ^{a}	0.00 ^{a}	0.01 ^{a}	0.00 ^{a}	0.00 ^{a}	0.00 ^{a}	0.00 ^{a}	0.80 ^{j}
	Up-Down	0.01 ^{a}	0.00 ^{a}	0.00 ^{a}	0.00 ^{a}	0.04 ^{ab}	0.00 ^{a}	0.00 ^{a}	0.00 ^{a}	0.02 ^{ab}	0.65 ^{cdefghi}

For the same column, mean values with different superscripts are significantly different ($p \leq 0.05$). Values are average citations over a 10 s oral processing period. A1 is a commercial reference. A2 and A3 are selected commercial porridge samples. Swallow (+) refers to easy to swallow.

3.2. Dynamic Oral Processing Trajectories for Selected Indigenous/Local CPs and a Commercial Porridge Reference

The correlation between the CPs and attribute changes over the 30 s OP duration (both Normal and Up-Down methods) was explored graphically via PCA product trajectory biplots on three dimensions (D1 to D3) (Figure 2).

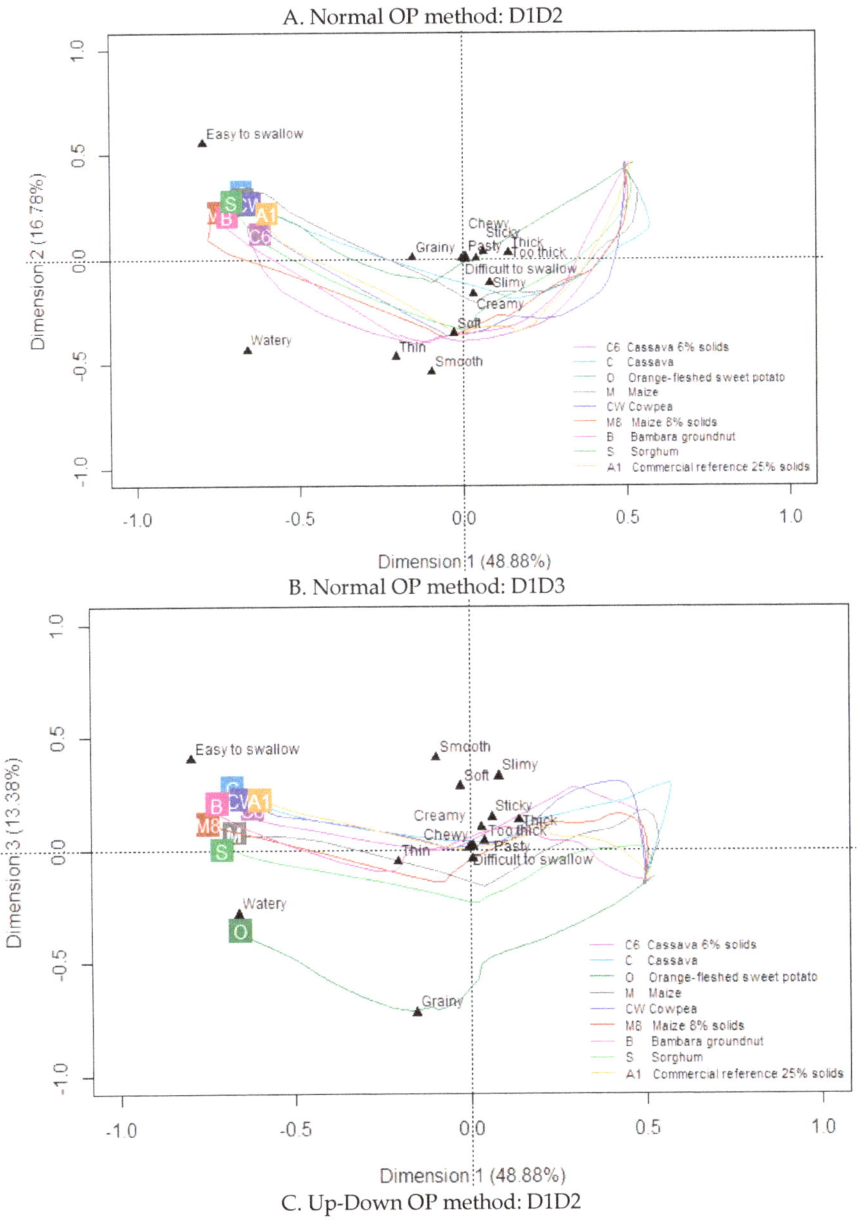

Figure 2. *Cont.*

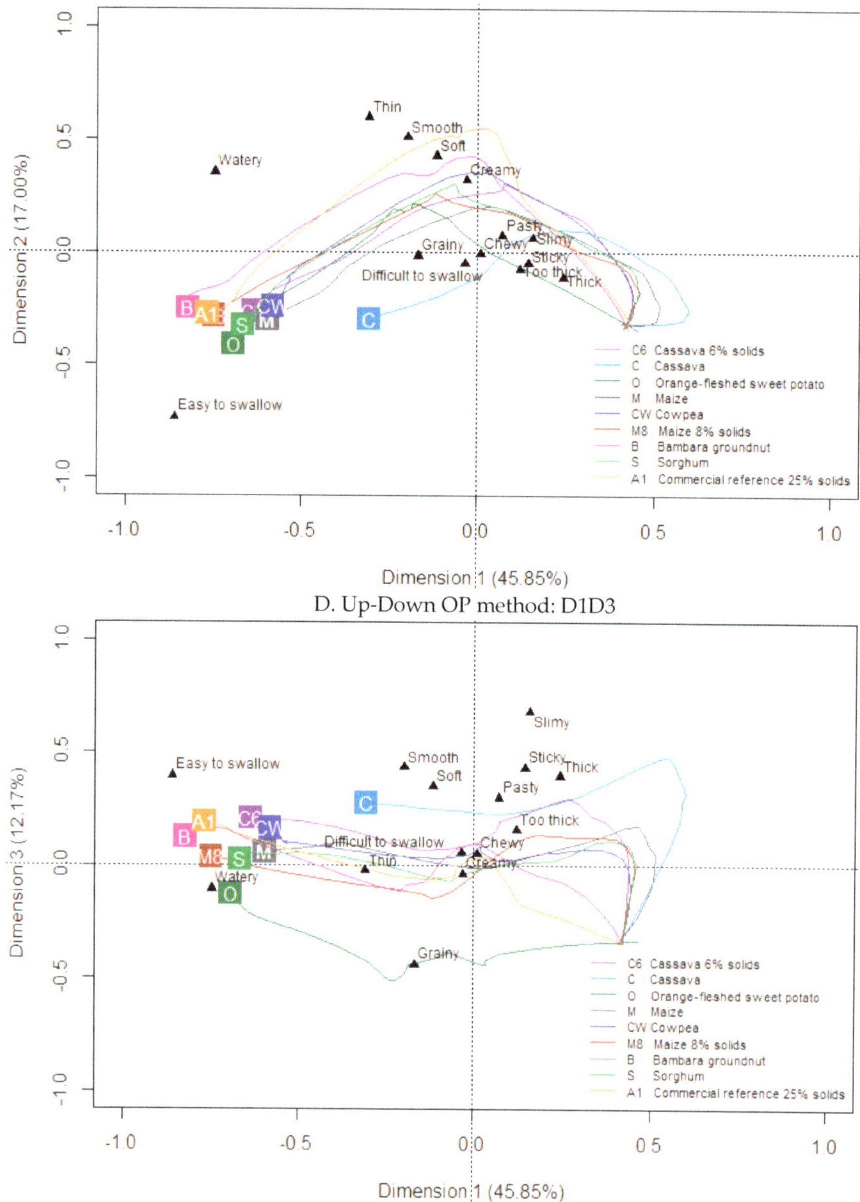

Figure 2. Smoothed Principal Component Analysis (PCA) product trajectory biplots show the evolution of attributes during oral processing for nine complimentary porridges (CPs) at 10% solids unless otherwise specified.

The first three dimensions (D1 to D3) in the normal OP method explained 79 % of the total inertia, while for the Up-Down OP method, these explained 75% of the model information. For both methods, during the evaluation, D1 (Figure 2A,C) explains the early CP differences due to attributes thick, slimy, sticky (and too thick in Up-Down method) in contrast to differences in watery and easy to swallow perceived towards the end of OP. Soft, grainy, smooth, and thin became relevant differentiation attributes midway in the evaluation. In both methods, D2 is associated with CP differences described as smooth, thin, watery, and creamy (Figure 2A,C). It is, however, clear that the temporal distribution and trajectory of CPs is different in D2 for the two OP methods. This implies that the CPs are perceived differently in time, when processed by Up-Down or Normal methods. With the Up-Down method (Figure 2C), the commercial porridge (A1) is clearly distinguished from the other CPs and, particularly, Cassava (C). With Normal OP (Figure 2A), the commercial porridge was not distinguished in D2 from most of the indigenous CPs, while OFSP (O), Cassava (C), and Maize porridges were more closely grouped. However, while all CPs were watery and easy to swallow, at the end of OP, in the Normal method, there was more variation with regards to these attributes amongst CPs in the Up-Down method (Figure 2C). Cassava was the least watery and least easy to swallow. Differences in the perception of the CPs as a function of the two OP methods are also explained in the D3 axes of the respective maps (Figure 2B,D). With the Normal method (Figure 2B), D3 separates CPs that are easy to swallow, smooth, soft, and slimy at the top from those that are grainy and to a lesser extent watery at the bottom of the plot (notably OFSP, O). With the Up-Down method (Figure 2D), slimy at the top is the main distinguishing attribute, particularly, contrasting the texture of mainly Cassava CPs to the grainy and watery character of OFSP (O) at the bottom of the plot.

4. Discussion

The early detection of viscosity-related attributes (too thick, thick, sticky, slimy, pasty) during OP of CPs was consistent with the generally known evolution of sensory perception during OP of semi-solid foods, as reported by [60]. The attributes thick, too thick, sticky, pasty and slimy, and less easy to swallow characterized the most common indigenous African CPs (Maize, Cassava, and Sorghum, 6–10% solids) early in OP. Similarly, tracking the changes in the oral texture of soft and semi-solid foods, [61] found that bulk-related attributes, such as thickness, dominate the initial phase of OP.

The characterization of Cassava, Maize, and Sorghum CPs (all 10% solids) as sticky, thick, too thick, slimy, and pasty, especially when eaten using the Up-Down method, has important implications for infant feeding. Children generally dislike sticky and slimy textures because of their lack of control over these texture attributes in the mouth [62]. A study [63] on the temporal texture quality of two soft cereal products of different composition (sponge-cake and brioche) for the elderly found that oral eating quality was negatively correlated with perceived stickiness and pastiness, but positively correlated with perceived easiness to eat, easy to chew, and easy to swallow the food bolus. The authors reported that sticky and pasty attributes were perceived and characterized as unpleasant [63]. According to [64], stickiness in the mouth and cohesiveness are the most important textural attributes to control in thin porridge. The developmental differences in oral physiology may impact the infants and young children's ability to orally process thick and sticky indigenous CPs. The age of a child impacts chewing and mastication abilities [65]. However, [14] observed that starting from 6 months onwards, acceptance of sticky textures was increasing with increasing infant age most likely due to familiarity. Eating skills develop gradually from sucking to munching and then chewing (a more complex rotary pattern) [15]. The capacity to transform food into a safely swallow-able bolus influences food acceptability and nutrient intake, yet it is severely limited in infants and young children [11]. For easy swallowing, the food bolus should be readily deformable and flowable [66]. The eruption of front teeth typically between 6 and 8 months and the distal ones from 12–24 months of age improves the ability to break down more challenging foods only later in infancy [67]. The change from the munching pattern to a more adapted, rotary chewing pattern occurs after at least 12 months of age [67,68]. This is because, in infants, milk and liquid foods are delivered directly to the posterior of the oral cavity during sucking and suckling- the optimal spot for swallow reflex initiation [67]. However, for handling solids, the infant must accept the food into the front of the mouth, masticate it, and then actively (with energy expenditure) use the tongue to transport the bolus to the posterior of the oral cavity for the swallow reflex to be triggered [69]. Mastication of solids-like, more challenging foods requires different chewing movements in bolus preparation [70], which are not present in early infants.

Cassava and Maize porridge (at 10% solids) were described as the least easy to swallow and still thick (21–30 s) in the Up-Down OP method; observations which were absent when these porridges were consumed in the Normal OP method. Although the oral physiology of infants is quite different from that of adults [71], in a study involving young adults (mean age 26.5 years) with a high number of functional dental units, and some elders (mean age 67.2 years) with varying numbers of opposing post-canine teeth pairs (i.e., functional units) it was found that the same amount of work is needed to transform food from its initial form to an easy to swallow bolus regardless of age. The lingual propulsive force in adults is thought to be the main driving force for bolus flow [72], making OP much easier in adults than in infants. Reduced dentition as is the case in infants compromises masticatory efficiency, as well as the tongue's capability in positioning food, and this reduces the efficiency of food oral break down [73,74]. It is acknowledged that young children and adults differ in their oral processing abilities, and that the adults' usage of different oral processing apparatus, saliva secretion, etc. cannot fully explain the texture perception and eating behavior of small children.

Cassava porridges (6–10% solids) were described as smooth during the initial OP phase and easy to swallow in the normal OP method, but thick, sticky, and less easy to swallow with the Up-Down OP method. The capacity of the Up-Down OP method to detect sensory differences at relatively high viscosity levels confirmed a proposition [75], that low shear viscosity is more relevant in differentiating thickness perception in fluid foods. The Up-Down method employed low shear viscosity, making handling and break down of thick and sticky porridges more problematic, while shear was higher with the Normal OP method, enabling greater intraoral food-mixing ability. In a research study on the dynamic texture break down of some soft foods (caramel), [76] noted that an increase in stickiness only led to an increase in the total amount of muscle used but to a slower masticatory process with larger opening and closing strokes. In infants, the more advanced motor skills for handling semisolid foods only appear between 9–12 months, followed by molars at 12–18 months [77]. As clinically noted, when children do not have the required OP abilities to break down foods, they often hold them in the mouth to soften with saliva and/or attempt to swallow the pieces whole, which increases the risk of choking [67,69,78]. The α-amylase and the lingual lipase enzymes in saliva are thought to digest starch and lipids respectively, partially contributing to a decrease in the perceived thickness of fluids [79]. However, in this study, saliva may not have influenced the differences in texture across OP methods, since the panel was constant.

Maize porridge (10% solids) in the Up-Down method was pasty midway during OP, while soft and easy to swallow by the Normal OP method at 8 % solids. The apparent more pleasant oral texture perception in the Normal OP method may be explained by the greater efficiency of use of dentition and other developed oral structures, such as the tongue, palate and jaws. Teeth action improves oro-tactile sensitivity and assists in cutting and grinding of more textured foods [80]. The tongue has a fundamental role in bolus containment and propulsion [81]. Fluid food thickness affects the chewing rate and muscular work [82]. A thick food bolus is difficult to deform and flow, as swallowing muscles may have to work much harder and much longer to generate enough oral pressure/stress to transform the bolus into appropriate flow-ability for swallowing [66]. The clear shift in oral texture from thick, sticky, and too thick, to thin and watery in the Normal OP method compared to the Up-Down method was probably due to a high shear rate in the Normal OP method because of a more complete (unrestricted) and effective use of the fully developed oral physiology. In a study on infant formulas, [83] reported a similar decrease in viscosity with increasing shear load (non-Newtonian shear thinning behavior). The oral shear rates in infants are extremely low due to the absence of lateral jaw movements. When a bolus is easily deformable and flowable, it can be swallowed comfortably with minimal oral effort. A limitation of the study is that the panel always applied the Up-Down method first followed by the Normal OP method, which could potentially affect results through possible order effects.

The relative increase in the initial perceptions of thick, too thick, sticky and pastiness of CPs with increasing solids content show a progression towards a more solid food state, which would be more challenging for infants to orally process. In infants whose OP is limited, this may lead to increased difficulty in food oral break down and oral manipulation. Food structure determines, to a large extent, how fast the food falls apart in the mouth, which influences the total chewing time and number of chews [84]. The impact of food composition on texture perception originate from differences in microstructural and physical-chemical properties [60]. The bulk volume of indigenous porridges results from gelatinized starch [85]. Consumer perception of the texture of porridge is influenced by processing technique and solids concentration [86]. For commercial CPs, processing steps, such as hydrolysis, dextrinization, and pre-gelatinization, contribute to pleasant oral sensory texture by breaking down complex and long chains of food biopolymers into short-chain molecules, lowering thickness and stickiness of porridges [83,87]. Foods are processed differently in the mouth, depending on their physical-chemical and mechanical properties [84]. As observed by [88], high bolus viscosity increases submental electromyographic (EMG) activity, indicating the use of additional muscular force during swallowing as viscosity increases. [89] reported that compromised OP capabilities (e.g., during transporting food to the mouth, opening and closing the mouth, and swallowing) is closely associated with a high level of eating difficulty, low energy intake, and malnutrition.

Food composition and structure also influence mastication (number of chews) and salivation [90]. Ref. [91], studying the sensory properties of a variety of foods showed that the stress applied during consumption depends on the viscosity of the food. Low viscosity foods were observed to be associated with minimum stress and increasing rate of deformation, while for high viscosity foods, the deformation rate was maintained as the stress was increased [84].

The effect of the OP method, porridge type, and the temporal nature of OP (i.e., the evolution of oral texture attributes over time) on the in-mouth texture perceptions of CP samples was demonstrated. Food texture perception is a highly dynamic process that depends on the constant manipulations and transformations of foods in the oral cavity [60]. In both OP methods, the initial texture of Maize, Cassava, and Sorghum CP (10%) was described as thick and/or too thick, sticky, slimy, progressing to soft, smooth, watery, then easy to swallow towards the end of OP. This is partly due to the shear-thinning behavior of the CPs, in addition to food oral breakdown during OP. At very low shear rates (zero-shear viscosity $\eta 0$), polymer suspensions are entangled many times, showing visco-elastic behavior arising from the balance between molecular disengagements and elastic recoil when an initial shear force is applied [35]. Each polymer chain assumes a spherical shape, entangled many times with neighboring macromolecules, leading to a higher viscosity at rest [36], often perceived as initial thickness and stickiness. When a shear load is applied during chewing, food molecules disentangle to a certain extent and gets aligned in the shear direction, and agglomerates disintegrate releasing bound liquid to flow again [35]. Together with the possible dilution and hydrolysis effects of salivary compenents, these events reduce porridge viscosity as OP progresses, which is perceived as a thin and watery consistency towards swallowing.

According to [92], the attribute "soft" when used to describe a food during OP implies pleasantness, and "easy to swallow" denotes a pleasant feeling as the bolus pass through the throat. "Viscous (thick)" and "sticky" imply unpleasantness of a material that adheres to or entangles on the eating utensils or teeth and is difficult to remove. OP was more complete in the Normal method as all samples achieved a watery and easy to swallow state, while in the Up-Down method, the samples achieved varying degrees of OP at the 30 s end-point, with Cassava (10% solids) the least easy to swallow. This correlated strongly with its perceived stickiness, pastiness, and sliminess prior to swallowing.

The reference porridge (A1) was characterized as creamy, smooth, and soft towards the end of OP. The feeling of creaminess is associated with the lubrication properties of the oil droplets between the tongue and the palate [84]. Smoothness is a complex tactile sensation implying the absence of graininess [93]. From an oral motor development perspective, optimal complementary foods should be well suited to infants' chewing and swallowing abilities in order to provide a pleasant early feeding experience. Yet that seems to not be the case with common indigenous and locally available CPs. An immediate effect of eating difficulty is reduced food intake, increasing the risk of malnutrition [94,95].

5. Conclusions

This study applied a trained adult sensory panel to gain insights into the temporal oral texture characteristics of indigenous porridges for infants and young children. The oral texture sensory perceptions of porridge samples were different depending on the OP method used. Indigenous CPs were thick, sticky, pasty, and slimy even at very low solids content, making the porridges potentially difficult to process, unpleasant, and not easy to swallow. The Up-Down OP method that mimics the restricted oral processing abilities of infants and young children leads to more enhanced perceptions of the thick, too thick, sticky, slimy, pasty, and difficulty to swallow attributes. This could ultimately limit food and nutrient intake, perpetuating protein and energy malnutrition in infants that rely on these food types. OFSP porridge had a satisfactory oral texture at its highest solids content, comparable to a commercial reference (A1). Parents and caregivers are advised to consider the use of OFSP flour in composite with a legume (e.g., Cowpea or Bambara) for the preparation of CPs with relatively high solids content, suitable oral texture, and nutritive quality. Simple traditional approaches for reducing the viscosity of indigenous CPs at home (e.g., malting, fermentation, souring) need consideration by primary caregivers. This study provides scientific insight for baby foods manufacturers on the OP characteristics of complementary foods for infants and young children in African communities. Smart tech innovations for processing indigenous flours to give CP an optimal oral texture at much higher solids content for improved infant nutrient intake, are required. Moreover, further research is needed to explore the dynamic sensory interplay between bolus properties of baby foods and the oro-tactile phenomena (tongue coordination, mastication, and lubrication) in infants and young children, which, at present, is not well understood.

Author Contributions: Conceptualization, J.M., H.D.K., and N.M.E.; formal analysis/Statistics, J.M. and T.C.; investigation, J.M.; resources, H.D.K. and N.M.E.; data curation, J.M., H.D.K.; writing—original draft preparation, J.M.; writing—review and editing, H.D.K., T.C., and N.M.E.; funding acquisition, J.M., H.D.K., and N.M.E.

Funding: This research is based on the research supported in part by the National Research Foundation (NRF) of South Africa, Department of Science and Technology DST/NRF Centre of Excellence in Food Security and the DST-CSIR Interbursary Support (IBS) Programme (Scholarship for J. Makame).

Acknowledgments: Marise Kinnear for technical support during the sensory research design; Hely Tuorila of the University of Helsinki, Finland for some constructive feedback of the first draft of the manuscript; John Castura from Compusense for technical advice on some study design and statistical analysis matters.

Conflicts of Interest: The authors declare no conflict of interest. The funders had no role in the design of the study; in the collection, analyses, or interpretation of data; in the writing of the manuscript, or in the decision to publish the results.

References

1. Chen, J. Food oral processing—A review. *Food Hydrocoll.* **2009**, *23*, 1–25. [CrossRef]
2. Stieger, M.; van de Velde, F. Microstructure, texture and oral processing: New ways to reduce sugar and salt in foods. *Curr. Opin. Colloid Interface Sci.* **2013**, *18*, 334–348. [CrossRef]
3. Aguayo-Mendoza, M.G.; Ketel, E.C.; van der Linden, E.; Forde, C.G.; Piqueras-Fiszman, B.; Stieger, M. Oral processing behavior of drinkable, spoonable and chewable foods is primarily determined by rheological and mechanical food properties. *Food Qual. Prefer.* **2019**, *71*, 87–95. [CrossRef]
4. Seidel, K.; Kahl, J.; Paoletti, F.; Birlouez, I.; Busscher, N.; Kretzschmar, U.; Särkkä-Tirkkonen, M.; Seljåsen, R.; Sinesio, F.; Torp, T. Quality assessment of baby food made of different pre-processed organic raw materials under industrial processing conditions. *J. Food Sci. Technol.* **2015**, *52*, 803–812. [CrossRef] [PubMed]
5. Breen, S.P.; Etter, N.M.; Ziegler, G.R.; Hayes, J.E. Oral somatosensory acuity is related to particle size perception in chocolate. *Sci. Rep.* **2019**, *9*, 7437. [CrossRef] [PubMed]
6. WHO/UNICEF. *Global Strategy for Infant and Young Child Feeding*; World Health Organization: Geneva, Switzerland, 2003.

7. Faber, M.; Laubscher, R.; Berti, C. Poor dietary diversity and low nutrient density of the complementary diet for 6-to 24-month-old children in urban and rural KwaZulu-Natal, South Africa. *Matern. Child Nutr.* **2016**, *12*, 528–545. [CrossRef] [PubMed]
8. Ratnayake, W.S.; Jackson, D.S. Gelatinization and solubility of corn starch during heating in excess water: New insights. *J. Agric. Food Chem.* **2006**, *54*, 3712–3716. [CrossRef] [PubMed]
9. Li, Y.; Shoemaker, C.F.; Ma, J.; Moon, K.J.; Zhong, F. Structure-viscosity relationships for starches from different rice varieties during heating. *Food Chem.* **2008**, *106*, 1105–1112. [CrossRef]
10. Amagloh, F.K.; Mutukumira, A.N.; Brough, L.; Weber, J.L.; Hardacre, A.; Coad, J. Carbohydrate composition, viscosity, solubility, and sensory acceptance of sweetpotato-and maize-based complementary foods. *Food Nutr. Res.* **2013**, *57*. [CrossRef]
11. Schwartz, C.; Vandenberghe-Descamps, M.; Sulmont-Rossé, C.; Tournier, C.; Feron, G. Behavioral and physiological determinants of food choice and consumption at sensitive periods of the life span, a focus on infants and elderly. *Innov. Food Sci. Emerg. Technol.* **2018**, *46*, 91–106. [CrossRef]
12. Akombi, B.J.; Agho, K.E.; Merom, D.; Renzaho, A.M.; Hall, J.J. Child malnutrition in sub-Saharan Africa: A meta-analysis of demographic and health surveys (2006–2016). *PLoS ONE* **2017**, *12*, e0177338. [CrossRef] [PubMed]
13. Dewey, K.G.; Brown, K.H. Update on technical issues concerning complementary feeding of young children in developing countries and implications for intervention programs. *Food Nutr. Bull.* **2003**, *24*, 5–28. [CrossRef] [PubMed]
14. Demonteil, L.; Tournier, C.; Marduel, A.; Dusoulier, M.; Weenen, H.; Nicklaus, S. Longitudinal study on acceptance of food textures between 6 and 18 months. *Food Qual. Prefer.* **2019**, *71*, 54–65. [CrossRef]
15. Nicklaus, S.; Demonteil, L.; Tournier, C. Modifying the texture of foods for infants and young children. In *Modifying Food Texture*; Chen, J., Rosenthal, A., Eds.; Elsevier: Amsterdam, The Netherlands, 2015; pp. 187–222.
16. Le Révérend, B.J.; Edelson, L.R.; Loret, C. Anatomical, functional, physiological and behavioural aspects of the development of mastication in early childhood. *Br. J. Nutr.* **2014**, *111*, 403–414. [CrossRef] [PubMed]
17. WHO. *Complementary Feeding: Report of the Global Consultation, and Summary of Guiding Principles for Complementary Feeding of the Breastfed Child*; World Health Organization: Geneva, Switzerland, 2003.
18. Balasubramanian, S.; Kaur, J.; Singh, D. Optimization of weaning mix based on malted and extruded pearl millet and barley. *J. Food Sci. Technol.* **2014**, *51*, 682–690. [CrossRef] [PubMed]
19. Brown, K.; Henretty, N.; Chary, A.; Webb, M.F.; Wehr, H.; Moore, J.; Baird, C.; Díaz, A.K.; Rohloff, P. Mixed-methods study identifies key strategies for improving infant and young child feeding practices in a highly stunted rural indigenous population in Guatemala. *Matern. Child Nutr.* **2016**, *12*, 262–277. [CrossRef] [PubMed]
20. Saitoh, E.; Shibata, S.; Matsuo, K.; Baba, M.; Fujii, W.; Palmer, J.B. Chewing and food consistency: Effects on bolus transport and swallow initiation. *Dysphagia* **2007**, *22*, 100–107. [CrossRef]
21. Alsanei, W.A.; Chen, J. Studies of the oral capabilities in relation to bolus manipulations and the ease of initiating bolus flow. *J. Texture Stud.* **2014**, *45*, 1–12. [CrossRef]
22. Steele, C.; Alsanei, W.; Ayanikalath, S.; Barbon, C.; Chen, J.; Cichero, J.; Coutts, K.; Dantas, R.; Duivestein, J.; Giosa, L. The influence of food texture and liquid consistency modification on swallowing physiology and function: A systematic review. *Dysphagia* **2015**, *30*, 2–26. [CrossRef]
23. Black, R.E.; Makrides, M.; Ong, K.K. *Complementary Feeding: Building the Foundations for a Healthy Life: 87th Nestlé Nutrition Institute Workshop, Singapore, May 2016*; Karger Medical and Scientific Publishers: Basel, Switzerland, 2017.
24. Plemmons, L.; Resurreccion, A. A warm-up sample improves reliability of responses in descriptive analysis. *J. Sens. Stud.* **1998**, *13*, 359–376. [CrossRef]
25. Guinard, J.-X. Sensory and consumer testing with children. *Trends Food Sci. Technol.* **2000**, *11*, 273–283. [CrossRef]
26. Haro-Vicente, J.; Bernal-Cava, M.; Lopez-Fernandez, A.; Ros-Berruezo, G.; Bodenstab, S.; Sanchez-Siles, L. Sensory acceptability of infant cereals with whole grain in infants and young children. *Nutrients* **2017**, *9*, 65. [CrossRef] [PubMed]
27. Kevin, K. You've come a long way, baby-food. *Food Process.* **1995**, *56*, 61–64.

28. Madrelle, J.; Lange, C.; Boutrolle, I.; Valade, O.; Weenen, H.; Monnery-Patris, S.; Issanchou, S.; Nicklaus, S. Development of a new in-home testing method to assess infant food liking. *Appetite* **2017**, *113*, 274–283. [CrossRef] [PubMed]
29. Longfier, L.; Soussignan, R.; Reissland, N.; Leconte, M.; Marret, S.; Schaal, B.; Mellier, D. Emotional expressiveness of 5–6 month-old infants born very premature versus full-term at initial exposure to weaning foods. *Appetite* **2016**, *107*, 494–500. [CrossRef] [PubMed]
30. Rombo, G.O.; Taylor, J.R.N.; Minnaar, A. Effect of irradiation, with and without cooking of maize and kidney bean flours, on porridge viscosity and in vitro starch digestibility. *J. Sci. Food Agric.* **2001**, *81*, 497–502. [CrossRef]
31. Thaoge, M.; Adams, M.; Sibara, M.; Watson, T.; Taylor, J.; Goyvaerts, E. Production of improved infant porridges from pearl millet using a lactic acid fermentation step and addition of sorghum malt to reduce viscosity of porridges with high protein, energy and solids (30%) content. *World J. Microbiol. Biotechnol.* **2003**, *19*, 305–310. [CrossRef]
32. Cichero, J.A. Unlocking opportunities in food design for infants, children, and the elderly: Understanding milestones in chewing and swallowing across the lifespan for new innovations. *J. Texture Stud.* **2017**, *48*, 271–279. [CrossRef] [PubMed]
33. Cook, D.J.; Hollowood, T.A.; Linforth, R.S.T.; Taylor, A.J. Correlating instrumental measurements of texture and flavour release with human perception. *Int. J. Food Sci. Technol.* **2005**, *40*, 631–641. [CrossRef]
34. Wool, R.P. Polymer entanglements. *Macromolecules* **1993**, *26*, 1564–1569. [CrossRef]
35. Mezger, T.G. *The Rheology Handbook*, 4th ed.; Vincentz Network: Hannover, Germany, 2014.
36. Gina. *Basics of Applied Rheology*; Anton Paar: Graz, Austria, 2016.
37. Vickers, Z.; Damodhar, H.; Grummer, C.; Mendenhall, H.; Banaszynski, K.; Hartel, R.; Hind, J.; Joyce, A.; Kaufman, A.; Robbins, J. Relationships among rheological, sensory texture, and swallowing pressure measurements of hydrocolloid-thickened fluids. *Dysphagia* **2015**, *30*, 702–713. [CrossRef] [PubMed]
38. Boesveldt, S.; Bobowski, N.; McCrickerd, K.; Maître, I.; Sulmont-Rossé, C.; Forde, C.G. The changing role of the senses in food choice and food intake across the lifespan. *Food Qual. Prefer.* **2018**, *68*, 80–89. [CrossRef]
39. Ndagire, C.T.; Muyonga, J.H.; Manju, R.; Nakimbugwe, D. Optimized formulation and processing protocol for a supplementary bean-based composite flour. *Food Sci. Nutr.* **2015**, *3*, 527–538. [CrossRef] [PubMed]
40. International Organization for Standardization. *Sensory Analysis: General Guidelines for the Selection, Training and Monitoring of Selected Assessors and Expert Sensory Assessors*, 1st ed.; International standard; ISO 8586; International Organization for Standardization: Geneva, Switzerland, 2012.
41. Castura, J.C.; Antúnez, L.; Giménez, A.; Ares, G. Temporal Check-All-That-Apply (TCATA): A novel dynamic method for characterizing products. *Food Qual. Prefer.* **2016**, *47*, 79–90. [CrossRef]
42. Ares, G.; Alcaire, F.; Antúnez, L.; Vidal, L.; Giménez, A.; Castura, J.C. Identification of drivers of (dis) liking based on dynamic sensory profiles: Comparison of Temporal Dominance of Sensations and Temporal Check-all-that-apply. *Food Res. Int.* **2017**, *92*, 79–87. [CrossRef] [PubMed]
43. Esmerino, E.A.; Castura, J.C.; Ferraz, J.P.; Filho, E.R.T.; Silva, R.; Cruz, A.G.; Freitas, M.Q.; Bolini, H.M. Dynamic profiling of different ready-to-drink fermented dairy products: A comparative study using Temporal Check-All-That-Apply (TCATA), Temporal Dominance of Sensations (TDS) and Progressive Profile (PP). *Food Res. Int.* **2017**, *101*, 249–258. [CrossRef] [PubMed]
44. Jaeger, S.R.; Alcaire, F.; Hunter, D.C.; Jin, D.; Castura, J.C.; Ares, G. Number of terms to use in temporal check-all-that-apply studies (TCATA and TCATA Fading) for sensory product characterization by consumers. *Food Qual. Prefer.* **2018**, *64*, 154–159. [CrossRef]
45. Baker, A.K.; Castura, J.C.; Ross, C.F. Temporal check-all-that-apply characterization of Syrah wine. *J. Food Sci.* **2016**, *81*, S1521–S1529. [CrossRef]
46. Varela, P.; Ares, G. *Novel Techniques in Sensory Characterization and Consumer Profiling*; CRC Press: Boca Raton, FL, USA, 2014.
47. Reyes, M.M.; Castura, J.C.; Hayes, J.E. Characterizing dynamic sensory properties of nutritive and nonnutritive sweeteners with temporal check-all-that-apply. *J. Sens. Stud.* **2017**, *32*, 1–25. [CrossRef]
48. Nguyen, P.T.; Kravchuk, O.; Bhandari, B.; Prakash, S. Effect of different hydrocolloids on texture, rheology, tribology and sensory perception of texture and mouthfeel of low-fat pot-set yoghurt. *Food Hydrocoll.* **2017**, *72*, 90–104. [CrossRef]

49. Dietitians Association of Australia; The Speech Pathology Association of Australia Limited. Texture-modified foods and thickened fluids as used for individuals with dysphagia: Australian standardised labels and definitions. *Nutr. Diet.* **2007**, *64*, S53–S76. [CrossRef]
50. Lazo, O.; Claret, A.; Guerrero, L. A comparison of two methods for generating descriptive attributes with trained assessors: Check-all-that-apply (CATA) vs. free choice profiling (FCP). *J. Sens. Stud.* **2016**, *31*, 163–176. [CrossRef]
51. Chambers, E., IV; Jenkins, A.; Garcia, J.M. Sensory texture analysis of thickened liquids during ingestion. *J. Texture Stud.* **2017**, *48*, 518–529. [CrossRef] [PubMed]
52. Meyners, M.; Castura, J.C. Randomization of CATA attributes: Should attribute lists be allocated to assessors or to samples? *Food Qual. Prefer.* **2016**, *48*, 210–215. [CrossRef]
53. McMahon, K.M.; Culver, C.; Castura, J.C.; Ross, C.F. Perception of carbonation in sparkling wines using descriptive analysis (DA) and temporal check-all-that-apply (TCATA). *Food Qual. Prefer.* **2017**, *59*, 14–26. [CrossRef]
54. R Core Team. *R: A Language and Environment for Statistical Computing*; R Foundation for Statistical Computing: Vienna, Austria, 2018; Available online: https://www.R-project.org/ (accessed on 16 February 2019).
55. Castura, J.C. tempR: Temporal Sensory Data Analysis. 2018. Available online: http://www.cran.r-project.org/package=tempR/ (accessed on 16 February 2019).
56. Meyners, M.; Castura, J.C. The analysis of temporal check-all-that-apply (TCATA) data. *Food Qual. Prefer.* **2018**, *67*, 67–76. [CrossRef]
57. XLStat. XLStat Help Documentation. 2018. Available online: https://help.xlstat.com/customer/en/portal/articles/2178395-download-the-xlstat-help-documentation (accessed on 27 May 2019).
58. Addinsoft. *XLSTAT Statistical and Data Analysis Solution*; Addinsoft: Long Island, NY, USA, 2019.
59. Marascuilo, L.A.; McSweeney, M. *Nonparametric and Distribution-Free Methods for the Social Sciences*; Brooks-Cole Publishing Co.: Stamford, CT, USA, 1977.
60. Devezeaux de Lavergne, M.; van de Velde, F.; Stieger, M. Bolus matters: The influence of food oral breakdown on dynamic texture perception. *Food Funct.* **2017**, *8*, 464–480. [CrossRef]
61. Scholten, E. Understanding perception of food types in terms of their structures: The missing links. *Food Funct.* **2017**, *8*, 462–463. [CrossRef]
62. Szczesniak, A.S. Texture is a sensory property. *Food Qual. Prefer.* **2002**, *13*, 215–225. [CrossRef]
63. Assad-Bustillos, M.; Tournier, C.; Septier, C.; della Valle, G.; Feron, G. Relationships of oral comfort perception and bolus properties in the elderly with salivary flow rate and oral health status for two soft cereal foods. *Food Res. Int.* **2019**, *118*, 13–21. [CrossRef]
64. Aboubacar, A.; Kirleis, A.W.; Oumarou, M. Important Sensory Attributes Affecting Consumer Acceptance of Sorghum Porridge in West Africa as Related to Quality Tests. *J. Cereal Sci.* **1999**, *30*, 217–225. [CrossRef]
65. Fucile, S.; Wright, P.M.; Chan, I.; Yee, S.; Langlais, M.E.; Gisel, E.G. Functional oral-motor skills: Do they change with age? *Dysphagia* **1998**, *13*, 195–201. [CrossRef] [PubMed]
66. Chen, J.; Lolivret, L. The determining role of bolus rheology in triggering a swallowing. *Food Hydrocoll.* **2011**, *25*, 325–332. [CrossRef]
67. Cichero, J. Introducing solid foods using baby-led weaning vs. spoon-feeding: A focus on oral development, nutrient intake and quality of research to bring balance to the debate. *Nutr. Bull.* **2016**, *41*, 72–77. [CrossRef]
68. Rudolph, C.D.; Link, D.T. Feeding disorders in infants and children. *Pediatr. Clin. N. Am.* **2002**, *49*, 97–112. [CrossRef]
69. Morris, S.E.; Klein, M.D.; Klein, D. *Pre-Feeding Skills: A Comprehensive Resource for Mealtime Development*; Academic Press: New York, NY, USA, 2001.
70. Koç, H.; Çakir, E.; Vinyard, C.; Essick, G.; Daubert, C.; Drake, M.; Osborne, J.; Foegeding, E. Adaptation of oral processing to the fracture properties of soft solids. *J. Texture Stud.* **2014**, *45*, 47–61. [CrossRef]
71. Kohyama, K.; Mioche, L.; Bourdio, P. Influence of age and dental status on chewing behaviour studied by EMG recordings during consumption of various food samples. *Gerodontology* **2003**, *20*, 15–23. [CrossRef]
72. Gilbert, R.J.; Napadow, V.J.; Gaige, T.A.; Wedeen, V.J. Anatomical basis of lingual hydrostatic deformation. *J. Exp. Biol.* **2007**, *210 Pt 23*, 4069–4082. [CrossRef]
73. Fontijn-Tekamp, F.A.; van der Bilt, A.; Abbink, J.H.; Bosman, F. Swallowing threshold and masticatory performance in dentate adults. *Physiol. Behav.* **2004**, *83*, 431–436. [CrossRef]

74. Ketel, E.C.; Aguayo-Mendoza, M.G.; de Wijk, R.A.; de Graaf, C.; Piqueras-Fiszman, B.; Stieger, M. Age, gender, ethnicity and eating capability influence oral processing behaviour of liquid, semi-solid and solid foods differently. *Food Res. Int.* **2019**, *119*, 143–151. [CrossRef]
75. He, Q.; Hort, J.; Wolf, B. Predicting sensory perceptions of thickened solutions based on rheological analysis. *Food Hydrocoll.* **2016**, *61*, 221e232. [CrossRef]
76. Cakir, E.; Koc, H.; Vinyard, C.J.; Essick, G.; Daubert, C.R.; Drake, M.; Foegeding, E.A. Evaluation of texture changes due to compositional differences using oral processing. *J. Texture Stud.* **2012**, *43*, 257–267. [CrossRef]
77. Campoy, C.; Campos, D.; Cerdó, T.; Diéguez, E.; García-Santos, J.A. Complementary Feeding in Developed Countries: The 3 Ws (When, What, and Why?). *Ann. Nutr. Metab.* **2018**, *73* (Suppl. S1), 27–36. [CrossRef] [PubMed]
78. Gisel, E.G. Effect of food texture on the development of chewing of children between six months and two years of age. *Dev. Med. Child Neurol.* **1991**, *33*, 69–79. [CrossRef] [PubMed]
79. De Wijk, R.; Terpstra, M.; Janssen, A.; Prinz, J. Perceived creaminess of semi-solid foods. *Trends Food Sci. Technol.* **2006**, *17*, 412–422. [CrossRef]
80. Delaney, A.L.; Arvedson, J.C. Development of swallowing and feeding: Prenatal through first year of life. *Dev. Disabil. Res. Rev.* **2008**, *14*, 105–117. [CrossRef]
81. Marconati, M.; Engmann, J.; Burbidge, A.; Mathieu, V.; Souchon, I.; Ramaioli, M. A review of the approaches to predict the ease of swallowing and post-swallow residues. *Trends Food Sci. Technol.* **2019**, *86*, 281–297. [CrossRef]
82. Van der Bilt, A.; Abbink, J. The influence of food consistency on chewing rate and muscular work. *Arch. Oral Biol.* **2017**, *83*, 105–110. [CrossRef]
83. Prakash, S.; Ma, Q.; Bhandari, B. Rheological behaviour of selected commercially available baby formulas in simulated human digestive system. *Food Res. Int.* **2014**, *64*, 889–895. [CrossRef]
84. Scholten, E. Composite foods: From structure to sensory perception. *Food Funct.* **2017**, *8*, 481–497. [CrossRef]
85. Nout, M.J.R.; Ngoddy, P.O. Technological aspects of preparing affordable fermented complementary foods. *Food Control* **1997**, *8*, 279–287. [CrossRef]
86. Muoki, N. Nutritional, Rheological and Sensory Properties of Extruded Cassava-Soy Complementary Porridges. Ph.D. Thesis, University of Pretoria, Pretoria, South Africa, 2013.
87. Nasirpour, A.; Scher, J.; Desobry, S. Baby Foods: Formulations and Interactions (A Review). *Crit. Rev. Food Sci. Nutr.* **2006**, *46*, 665–681. [CrossRef] [PubMed]
88. Smith, C.H.; Logemann, J.A.; Burghardt, W.R.; Carrell, T.D.; Zecker, S.G. Oral sensory discrimination of fluid viscosity. *Dysphagia* **1997**, *12*, 68–73. [CrossRef] [PubMed]
89. Laguna, L.; Chen, J. The eating capability: Constituents and assessments. *Food Qual. Prefer.* **2016**, *48*, 345–358. [CrossRef]
90. Witt, T.; Stokes, J.R. Physics of food structure breakdown and bolus formation during oral processing of hard and soft solids. *Curr. Opin. Food Sci.* **2015**, *3*, 110–117. [CrossRef]
91. Shama, F.; Sherman, P. Identification of Stimuli controlling the Sensory evaluation of Viscosity II: Oral Methods. *J. Texture Stud.* **1973**, *4*. [CrossRef]
92. Hayakawa, F.; Kazami, Y.; Nishinari, K.; Ioku, K.; Akuzawa, S.; Yamano, Y.; Baba, Y.; Kohyama, K. Classification of J apanese Texture Terms. *J. Texture Stud.* **2013**, *44*, 140–159. [CrossRef]
93. Cai, H.; Li, Y.; Chen, J. Rheology and tribology study of the sensory perception of oral care products. *Biotribology* **2017**, *10*, 17–25. [CrossRef]
94. McLaren, S.; Dickerson, J. Measurement of eating disability in an acute stroke population. *Clin. Eff. Nurs.* **2000**, *4*, 109–120. [CrossRef]
95. Tamine, K.; Ono, T.; Hori, K.; Kondoh, J.; Hamanaka, S.; Maeda, Y. Age-related changes in tongue pressure during swallowing. *J. Dent. Res.* **2010**, *89*, 1097–1101. [CrossRef]

MDPI
St. Alban-Anlage 66
4052 Basel
Switzerland
Tel. +41 61 683 77 34
Fax +41 61 302 89 18
www.mdpi.com

Foods Editorial Office
E-mail: foods@mdpi.com
www.mdpi.com/journal/foods

www.ingramcontent.com/pod-product-compliance
Lightning Source LLC
LaVergne TN
LVHW070621170726
843515LV00004B/147